PAT REMLER

HANNAH

About the Authors

Bob Brier (left) is a world-famous Egyptologist who has conducted research on pyramids and tombs in fifteen countries. A senior research fellow at the C. W. Post campus of Long Island University, he is the author of seven books, including *The Murder of Tutankhamen*, and he has hosted the Great Egyptian series for the Learning Channel.

Jean-Pierre Houdin (right) left his Parisian architecture firm in 1999 to devote himself to solving the mystery of the Great Pyramid. He has been awarded the Montgolfier Prize for his research.

ALSO BY BOB BRIER

The Murder of Tutankhamen

Ancient Egyptian Magic

Egyptian Mummies: Unraveling the Secrets of an Ancient Art

How One Man's

Obsession Led to the Solution of

Ancient Egypt's Greatest Mystery

Smithsonian Books

HARPER

NEW YORK • LONDON • TORONTO • SYDNEY

THE

Secret

OF THE

Great Pyramid

Bob Brier

and

Jean-Pierre Houdin

HARPER

A hardcover edition of this book was published in 2008 by Smithsonian Books in association with HarperCollins Publishers.

FIRST HARPER PAPERBACK PUBLISHED 2009.

Designed by Sunil Manchikanti

Library of Congress Cataloging-in-Publication Data is available upon request.

ISBN 978-0-06-165553-1

09 10 11 12 13 OV/RRD 10 9 8 7 6 5 4 3 2 1

Dedication

I dedicate this book to my two closest friends: the video artist Bulle Plexiglass* (my wife Michelle) and Henri, my father, with whom I've all my life had a very strong relationship and a close complicity.

I want to thank Bulle for always pushing me to go to the essential in life, to live every day full-time with ethics and passion. That vision of life led us, in the late nineties, to take a sabbatical year in New York looking for new ideas. Thanks to that break—our break with routine—I came back to Paris with new tools (digital 3-D and Internet), ready for what I call my third life—a totally unexpected one, focusing on one of the last mysteries on earth: trying to resolve how the Great Pyramid was built! In a few words: the Man in Black would have never existed without Bulle.

I want to thank Henri for having ignited, almost ten years ago, my passion for the Pyramid of Khufu with an idea of genius: that true great pyramids were built not from the outside but from within, a breakthrough concept relative to what has been thought for two hundred years. This is the idea that I unconsciously was looking for for my third life. And I'm proud that our Khufu Adventure has kept us close to each other. Henri recently turned eighty-five and is still "young and active"; I have the feeling that this Khufu Adventure is no little cause for that.

I want also to pay tribute to Renée, my late mother, who was always very anxious about the future of her son and daughter-in-law, concerned about the difficult period Bulle and I went through financially. Sadly, she passed away three years ago after a long and terrible illness. She was no longer with us when the "Khufu Story" started to become recognized and respected.

To Brigitte

—J-P.H.

*www.bulleplexiglass.net

Contents

Acknowledgments

Nearly ten years ago, when my father and I started our amazing adventure about Khufu, we were just the two of us. At the time of the publication of this book, we are no longer alone: hundreds of smart, skilled, and friendly people joined us throughout these long years. The tiny stream from the early times became a large river, still growing more and more, day after day. It would be quite impossible to thank by name each of them in these acknowledgments, but I want them to know how deep my gratitude is for their support, help, and advice.

I want to thank all the civil engineering experts from the CNISF (Conseil National des Ingénieurs et des Scientifiques de France) who were, from the end of 1999, the first interested in our studies. Their knowledge and competence in the construction field have always been very useful for us. Later, they formed the backbone of the ACGP (Association Construire la Grande Pyramide; www.construire-la-grande-pyramide.fr), an association set up in 2003 to financially support the development of the theory and to help set up a scientific survey on the Giza Plateau.

I am grateful to the founding members of the ACGP, among them Jean Billard, François Levieux, Jean-Louis Simonneau, Paul Allard,

Bernard Marrey, Paul Lemoine, Daniel Solvet, Charles Bambade, Dominique Ferré, Georges Rème, Emilienne Dubois, and Ruth Schumann-Antelme. Many others joined the ACGP, mainly as honorary members because of their valuable involvement as scientists, Egyptologists, engineers, personalities, or simply people passionate about ancient Egypt.

My thanks go to my Egyptian friends, among them Mourad M. Bakhoum, Farid Atiya, Hany Helal, Naheed Abdel Reheem, Brigitte Boulad, Sherine Mishriki, Hassan Benham, Mounir Neamatalla, Taha Abdallah, Mahmoud Ismail, Essam El Maghraby, Atef Moukhtar, Waffiq Shamma, Nadia Fanous, Sayed Kotb, and the staff of the Victoria Hotel in Cairo.

My thanks go to my French friends, among them Denis Denoël, Marc Buonomo, Pierre Grussenmeyer, François Schlosser, Jean Carayon, Hubert Labonne, Jean Berthier, Raoul Jahan, Hervé Piquet, Raphaël Thierry, Albert Ranson, Jean-Jacques Urban-Galindo, Pierre Deletie (deceased), Hui Duong Bui, Guy Delbrel, Dominique Gimet, François de Closets, and . . . Bernard, my brother.

My thanks go to my other non-French friends, among them Craig B. Smith, Bob Goldberg, Roman Golicz, Jon Bodsworth, Jack Baklayan, Jeffrey Kearney, Lionel Woog, Mark Rose, Norman Stockle, Jack Scaparro, Pat Remler, and Bob Brier.

My thanks go to the following companies that have supported the Khufu Adventure since 2004 through financial or technical donations to the ACGP: Eiffel (Jacques Huillard), Dassault Systèmes (Bernard Charlès); Thales Group (Thierry Brizard); Schneider Electric Egypte, (Frédéric Abbal, Emmanuel Lemasson); Gaz de France Egypte (Patrick Longueville, Jean-Louis Chenel); Air France (François Brousse); Jacquet SA (Christian Jacquet), GPI (Michel Gergonne); Borifer FIB (Jean-François Bordenave); Enerpac France (Guillaume Butty); Groupe Pyramides Egypte (Bruno Neyret); Arab Consulting Engineers (Mourad M. Bakhoum); and Farid Atiya Press (Farid Atiya). Without their support, the Khufu Adventure would have been long dead.

My thanks also for the Egyptologists Audran Labrousse, Dieter Arnold, and Rainer Stadelmann, who simply took the time to listen to me, even though I don't belong to the Egyptology community. The interest they showed and their advice are those of respectable people.

I wish to include a special acknowledgment to Mehdi Tayoubi, Richard Breitner, Fabien Barati, Emmanuel Guerriero, and the whole

"Khufu Revealed" team (www.3ds.com/khufu) at Dassault Systèmes and Emissive. We have already spent three years together on Khufu's Pyramid. Something tells me that we have a lot of work still ahead, and that one day in the near future we will be able to tell more about this amazing pyramid.

Special thanks to my friend Laurent Chapus.

I hope those I haven't named above will forgive me, but I want them to be sure that I think about them as much as those I have.

Bob, Pat . . . thank you for all.

J-P.H.

There are many people to thank and on the top of my list is Jean-Pierre Houdin. Helping him with *Secret of the Great Pyramid* has been a wonderful intellectual adventure. I haven't learned so much from one person since graduate school! I asked so many questions about the Great Pyramid that we now both laugh when I say, "I have a question." Jean-Pierre, I hope you get to take your walk up the ramp.

On a more earthly level I am indebted to my agent, Liza Dawson. More than just an agent, Liza is a superb editor who always pushes me to say more and do better. When Liza was satisfied that we had a book, she placed me in the capable hands of Elisabeth Dyssegaard, my editor at Smithsonian Books. Elisabeth understood the project right from the beginning, guided the book, improved my writing, and even made the endeavor fun. Her assistant, Kate Antony, somehow made sense of all Jean-Pierre's diagrams and illustrations and was midwife at the birth of order out of chaos.

As in all the books I have written, my wife, Pat Remler, not only forced me to clarify my murky prose, but also provided crucial photos. Finally, I want to thank yet another unofficial editor, Judith Turner, whose architectural knowledge saved me time and again.

Thank you all!

B.B.

The Man in Black

Almost all Egyptologists receive mail from strangers. Sometimes it is from reincarnated pharaohs; sometimes it is from prospective tourists who want to know if they can drink the water. (You can't.) Because my specialty is mummies, I receive hundreds of letters and e-mails from sixth graders who have been studying Egypt in school and want to mummify their recently deceased parakeets. About twice a year I receive offers from people who want to donate their bodies for mummification when they die. To these good folks I reply that I did that once, as a research project, and have now moved on.

No matter what our Egyptological specialty, we all receive communications from retired engineers with theories of how the pyramids were built. Usually there is an obvious flaw that even I can spot. On June 16, 2003 I received an e-mail from a French architect, Jean-Pierre Houdin, who had his own theory of how the Great Pyramid of Giza had been built. My friend Jack Josephson, an art historian who also has a degree in engineering, had suggested he contact me. Jack is a no-nonsense type of guy; I knew that if he told Houdin to contact me, it would be worthwhile to meet him. So I invited Jean-Pierre for dinner. Also coming was my friend Armand, an engineer who had been to

Egypt, and his wife, as well as another friend, Jack Scaparro, who was working on a novel set in Egypt. My Egyptologist wife, Pat, rounded out a receptive audience of five.

Jean-Pierre arrived at precisely 4:00 P.M., as suggested. We wanted time to hear and discuss his theory before dinner at seven, but would soon discover that three hours was not nearly enough time. This was not your ordinary pyramid theory. Our guest was dressed all in black, including a fashionable black leather jacket—in New York in June. A well-manicured gray-haired man of fifty, he had a pleasant smile and spoke heavily accented but good English. We were soon gathered around the coffee table as Jean-Pierre set up his laptop. He explained that he had given up his architectural practice five years ago so he could devote himself to the puzzle of how the Great Pyramid was built. Working out of his Paris apartment, he spent six or seven hours a day creating elaborate 3-D computer simulations of the interior and exterior of the Great Pyramid. As his computer models progressed he became more and more obsessed with the Pyramid, until it was all he thought about.

Jean-Pierre's interest in the Great Pyramid began in 1999 when his father, an engineer, watched a television documentary on the pyramids and realized the program's presentation of how the Great Pyramid was built was all wrong. He had another idea of how the huge blocks were raised to the top, a revolutionary idea, different from anything anyone had ever proposed, so he called his son Jean-Pierre and laid it out.

The father-son team was ideal to tackle this mystery. Henri Houdin had earned a PhD in engineering from Paris's prestigious École des Arts et Métiers. In 1950, as a twenty-seven-year-old engineer, he was sent to Ivory Coast to build their infrastructure. When he arrived there were eight kilometers of paved roads; when he left there were highways, bridges, and power plants. For decades Jean-Pierre had designed houses and office buildings; he knew about planning big projects. The two were equipped to answer the question how the Great Pyramid was built, but it would not be easy. Eventually the search for the answer would take over both their lives.

While his laptop booted up, Jean-Pierre explained how his computer graphics helped him understand the interior and exterior of the Great Pyramid. With new, sophisticated software developed for architects, he created 3-D images of the chambers inside the Great Pyramid. Then, on his computer screen, he could rotate the images to see the spatial relationships between the rooms—what features were on the same level, what parts had to have been built first, where the largest stones in the Pyramid were placed.

As he clicked the keypad, beautiful diagrams of the Pyramid appeared and we realized we were in the presence of a man who knew the Great Pyramid intimately. He explained why some blocks in the Pyramid were limestone and others granite; why the patterns of stone in some walls were different from others. I have friends who are pyramid experts, but I had never heard anything from them as detailed as Jean-Pierre's explanation. I was astounded by the quality of his graphics. Little figures hauled blocks up inclined ramps and put them in place with ingenious lifting devices. He even had topographical maps of the Giza Plateau to show how the architects of the Great Pyramid took advantage of the natural contours of the land to move huge blocks of stone. The images weren't just informative, they were beautiful. I had just completed a high-budget documentary for The Learning Channel on pyramids around the world. We'd spent thousands and thousands of dollars on our graphics, and Jean-Pierre's were better!

Jean-Pierre explained the difficulties with the two competing theories of how the blocks in the Pyramid were raised to the top. The single ramp theory, so often shown in television documentaries, could be easily discredited. The basic idea is that blocks were hauled up a long ramp constructed against one of the sides of the Pyramid. As the Pyramid grew, the ramp was raised and extended. The problem is that to keep the slope gentle enough so men could haul blocks, the ramp would have to be a mile long. If the Pyramid were being built on the site of New York's Empire State Building, the ramp would extend all the way into Central Park, about twenty-five city blocks. Building just the ramp would have taken thousands of men decades. Also, there would have been a tremendous amount of debris from such a ramp, and rubble doesn't just disintegrate in the desert; but huge piles of rubble have never been found. Perhaps most damaging to the single ramp theory is the fact that there is practically no place to put such a long ramp on the Giza Plateau.

The second theory fared no better in Jean-Pierre's analysis. It posited that a ramp had corkscrewed around the outside of the Pyramid itself, like a road winding around a mountain. This solves the no-space problem. But this theory has a fatal flaw as well. The Pyramid has four corners, and as the Pyramid grew, the architects had to constantly sight along those corners to make sure the edges were straight and thus ensure that they would meet at a perfect point at the top. But a ramp corkscrewing up the outside would have obscured these sight lines. Thus this too couldn't be how the ancient Egyptians raised the blocks to the top.[1]

Jean-Pierre showed us graphics of what the Pyramid looked like, year by year, as it was being built. And then he sprang his theory. He claimed that *inside* the Great Pyramid was a mile-long ramp corkscrewing up to the top, that had remained undetected for 4,500 years! We were astonished. The theory was so radical, so different from anything ever imagined, that it seemed impossible. But as the parade of graphics continued on the computer screen and Jean-Pierre explained the details, it seemed more and more credible, even probable. Here was a solution that answered the questions Egyptologists had been asking for decades. Somehow the centuries-old mystery of the Great Pyramid had been solved by this intense Frenchman dressed in black. I had the feeling that this theory just might be one of the great moments in the intellectual history of the world.

We were all overwhelmed by Jean-Pierre's step-by-step explanation of the new theory, but the Jean-Pierre Pyramid Show was just beginning. For two more hours, a parade of graphics and explanations of Pyramid minutiae appeared on his laptop. We saw what the Pyramid looked like after five years, seven years, ten years, twenty years, until it seemed as if we were watching it rise block by block. There are three chambers inside the Great Pyramid, and Jean-Pierre gave us the construction details of all three. Two of the rooms—the King's Chamber and the Queen's Chamber—were clearly intended for burials. They are rectangular rooms where you can put a sarcophagus, statues, funerary furniture, and anything else needed for the next world. The great puzzle is the mysterious Grand Gallery. It doesn't make any architectural sense. It is a long hallway sloping upward inside the Pyramid, leading to the King's Chamber. But why the twenty-eight-foot ceiling? And why line the side walls with low stone benches with strange grooves carved in them? Our guest in black had an answer for every question.

After three hours Jean-Pierre was still just warming up, but he had lost his audience; we were all brain-dead. Jean-Pierre had no idea that much of what he had thrown at us had simply bounced off. None of us was nearly so familiar with the Pyramid as he was; there were concepts we just couldn't wrap our nonarchitect brains around. But we were all impressed. Jean-Pierre could visualize the Great Pyramid in three dimensions in a way no one had ever imagined possible. The man in black was amazing.

Feeding Jean-Pierre was the only way I could think of to slow him down. Still, he continued his Pyramid seminar, laying out the physical evidence to support his theory and so the discussion continued over pasta. By dessert the radical new idea was looking better and better. I asked Jean-Pierre what most impressed him when he first saw the Great Pyramid. "Oh, I've never been there," he answered nonchalantly. The man who had given up his career as an architect and devoted five years of his life unraveling the secrets of the Great Pyramid had never seen it—and didn't want to! When I pressed Jean-Pierre about visiting the Pyramid, he wasn't interested. "Oh, I know what it looks like," he said, and added some polite response like, "Oh yes, I must do that." Clearly, for him the Great Pyramid was an abstract intellectual puzzle to be solved, not a monument to be visited. Perhaps he was afraid he would be disappointed; perhaps he feared something he might see would disprove his theory and the search would be over.

As the evening wore on the question became, what to do next? Here was a man with what could be the most important archaeological discovery of the century. For five years he had devoted his life to solving an incredibly difficult intellectual puzzle. It was one of those rare times when obsession seems to have produced something positive—indeed, something wonderful, so wonderful that I found myself being drawn into the obsession. I wanted to be part of it; I wanted to be there when Jean-Pierre found his internal ramp.

Birth of the Pyramid

Of the seven wonders of the ancient world, only the pyramids still stand. Virtually nothing remains of the Hanging Gardens of Babylon; of the Colossus of Rhodes only an occasional fragment turns up in a fisherman's net, and nonhistorians would find it difficult to even name the remaining four (the Statue of Zeus at Olympia, the Temple of Artemis at Ephesus, the Lighthouse at Alexandria, and the Mausoleum at Halicarnassus). Each year the pyramids are visited by millions of tourists who stare in wonder at the achievement of ancient man. When you stand in front of the Great Pyramid you can almost *feel* its massive weight.

One hundred and seven pyramids are scattered throughout Egypt, but the pyramids on the Giza Plateau are the famous ones. They are the largest, best preserved, and most visited of all the pyramids. The Great Pyramid is only a few feet taller than the one next to it, but because of its unique maze of internal rooms and passages, it is the one with all the legends. The Pyramid's entrance remained unknown until the ninth century when the caliph Al Mamun tunneled deep into its core and hit a passageway. By torchlight his men followed the passageway upward until they discovered the Queen's Chamber, a small rectangular room with twelve huge limestone rafters forming the ceiling. To their disappointment, the room was empty, but still in search of treasure, they continued upward through the Grand Gallery to the King's Chamber, where they found only an empty sarcophagus; all the treasure had been carried off centuries before. Still, the adventures of Al Mamun's band of would-be tomb robbers became one of the 1001 tales in the *Arabian Nights*, that says Al Mamun found weapons of metal that would not rust and glass that would not break.[2] Modern science has studied, mapped, and photographed the Great Pyramid more than any other pyramid, yet it still harbors mysteries. There are rooms whose purposes are unknown, and only recently a robotic camera sent up the air shafts in the Queen's Chamber discovered tiny doors with copper handles. What lies behind these doors is still unknown.[3]

Of all these mysteries, the biggest is how the Great Pyramid was built. The Egyptians were a nation of accountants. They recorded everything—how many of the enemy they killed in battle, the names of the pharaoh's children, even cake recipes. We have medical papyri, short stories, laundry lists, reports of expeditions—but not a single document recording how the pyramids were built. I can tell you the names of the two horses that pulled Ramses the Great's chariot at the Battle of Kadesh (Mut Is Content and Victory in Thebes) but I can't tell you how the Egyptians hauled those huge blocks up the Great Pyramid or how many men worked on the Pyramid at any one time. The ancient Egyptian records are silent when it comes to the greatest building project in the history of man. Because of this, wild theories have always abounded.

When Menachem Begin, Israel's prime minister, visited the pyramids, he proudly proclaimed that his ancestors built them. The prime minister had bought into one of the many pyramid myths. When Begin's ancestors were in bondage in Egypt, the pyramids were already ancient. The Old Testament never mentions the pyramids. It does, however, say that the Israelites built the store cities of Ramses and Pithum, which places the Israelites sojourn in Egypt during the reign of Ramses the Great—a thousand years after the Great Pyramid was built.[4] Hebrew slaves hauling blocks is pure Hollywood. The pyramids were built by free labor—Egyptian construction workers who were paid for their services—but slaves building the Pyramid isn't the only Biblical nonsense connected with the Pyramid. There's a story in the Bible that has nothing to do with the pyramids, but in the eighteenth century somehow it was linked to them.

Remember the Biblical story of Joseph and how he interprets the pharaoh's dream of seven lean cows devouring seven fat cows?[5] Joseph explains that there will be seven prosperous years for Egypt, but they will be followed by seven years of famine. Based on Joseph's dream interpretation, grain is stored during the prosperous years so Egypt will not starve during the lean years. Where was all that grain stored? You guessed it—in the Great Pyramid. One of the earliest theories of the function of the pyramids was that they were "Joseph's granaries."

There are so many things fundamentally wrong with this theory that it could only have been invented by someone who had never seen the pyramids. Think about it. The pyramid would have to have been built in less than seven years in order to store the grain for the lean years. It might be argued that the pyramids were built as granaries before Joseph arrived on the scene; that they were always Egypt's way of storing grain. But we still have the problem of where to put the grain inside the pyramid. It's mostly solid! Even if the few interior rooms inside the pyramid were used to store the grain, that would entail it being carried through narrow, dark passages over slippery inclines and up to a great height inside the pyramid. No, whoever first concocted the Joseph's granary theory never saw the pyramids.

Prior to the eighteenth century, you could get away with this kind of theory because practically no one outside of Egypt had seen the pyramids. Most of the early published illustrations of the Giza pyramids were drawn by artists who had heard about them but certainly not seen them. They had been told that they were on a plateau, and were pointy, so they drew what they thought they looked like. Often they were too steep and too numerous. Finally, in the seventeenth century, when adventurous travelers began visiting Egypt, the theories about the Great Pyramid began to be based on observation, but that still didn't mean they were sensible.

One of the most common misconceptions about the Great Pyramid is that encoded in its measurements are secrets of the universe. In 1620, John Greaves, a young Oxford astronomer, visited the Pyramid.[6] Like many before him, he believed the Egyptians were an advanced civilization whose knowledge exceeded his. Greaves believed that the Egyptians knew the exact circumference of the earth, and was convinced

Early depictions of the pyramids of Giza were often drawn by artists who had never been to Egypt.

that they had built this calculation into the dimensions of the Pyramid. In preparation for his visit to Giza, Greaves measured ancient monuments throughout Europe, trying to determine units of measure. He became familiar with the Roman foot, the English foot, and the Greek foot, and all kinds of other measures. Armed with precisely made brass measuring rods, he sailed for Egypt to conduct the first scientific survey of the Pyramid.

Greaves's exploration of the Great Pyramid was quite different from what the modern tourist experiences. There were no lights. Greaves had to make his way by torchlight, passing through swarms of bats and more than once nearly passing out from the stench of the bat guano. When Greaves ascended the Grand Gallery to reach the King's Chamber, there was no wood planking over which tourists now walk. Nor were there handrails, so in the near dark, fighting bats and slippery inclines, Greaves worked his way into the bowels of the Pyramid. Through it all, dragging his measuring rods, he recorded the dimensions of the Pyramid as precisely as possible. In addition to his measurements, Greaves discovered a well-like chamber at the base of the Grand Gallery that was dug to provide air for the workers excavating the descending passageway.

When he returned to London, he published his results in a small book titled *Pyramidographia*, but it was all for naught. His measurements were inaccurate. In Greaves's day, the base of the Pyramid was covered with a thirty-foot-high pile of rubble that obliterated its corners. Greaves was forced to estimate the base of the Pyramid and was off by more than fifty feet. To determine the height, Greaves climbed the Pyramid, counting the courses as he went. Estimating the number covered with rubble, he came to a total of 297. Again he was off: his estimate of the height (499 feet) was too high.

Sometimes it wasn't just a matter of inaccurate measurements. Greaves was overwhelmed by the entire experience and got even the most basic things wrong. He says the burial chamber's walls are made of six levels of stone; it's five. Still, this was a pioneering attempt. Two hundred years later another astronomer would undertake a survey the Great Pyramid with greater accuracy.

In March 1864, Piazzi Smyth, Astronomer Royal of Scotland, sailed for Egypt to conduct the most detailed measurement ever of the Great Pyramid. He believed that the Pyramid's dimensions were the key to

Biblical revelations, so it was crucial for him to know exactly how tall, how wide, how deep the Pyramid and all its chambers were. Although the Pyramid predated Christianity by more than 2,000 years, Smyth was convinced it was actually a Christian monument. "The Great Pyramid was yet prophetically intended—by inspiration afforded to the architect from the one and only living God, who rules in heaven and announced vengeance against the sculptured idols of Egypt."[7]

Smyth threw himself into the task of surveying the Pyramid with religious abandon, spending money he didn't have to manufacture precise instruments for his survey. He even invented a miniature eight-inch camera with photographic plates about the size of a modern slide, so he could photograph in the smallest of crevices. Thus when in March of 1864 the Astronomer Royal and his devoted wife sailed for Egypt, with them went hundreds of pounds of carefully crated equipment they hoped would lead to revelations left by God that were encoded in the Pyramid. They arrived in Alexandria, made the daylong journey to Cairo, and soon took up residence in one of the tombs near the Pyramid.

Smyth believed that the unit of measurement used by the Pyramid's builders was something he called "the sacred cubit." If he could determine exactly how large the sacred cubit was, then he could determine how many cubits for the height of the Pyramid, how many cubits in the sides of the burial chamber, and so on. Then, once he had those dimensions, he could deduce the secret message they contained. For months Smyth crawled in and out of the Pyramid with his specially crafted measuring rods, "inclinators," and cameras. He did indeed conduct the most precise survey of the Pyramid ever, and concluded that coded into the Pyramid's measurements were the exact size, shape, and motions of the earth.[8] He concluded that the unit used in the construction of the Pyramid was also used in the construction of Solomon's Temple and Noah's Ark. From his measurements of the burial chamber and the empty sarcophagus in it, Smyth deduced the earth's density. The expedition was a remarkable combination of exacting science and delusion.

After Smyth returned home he published *Life and Work at the Great Pyramid* (1867). It was universally rejected by the scientific community as the ravings of a religious fanatic. For years Smyth churned out revised editions of his book and never gave up the theory. If the Astronomer Royal of Scotland formulated such theories, is it any wonder that some people today believe the Pyramid was built by aliens?

A 1970s fad invested the pyramid shape with magical properties. Place a dull razor blade inside a cardboard pyramid and it will sharpen; put meat inside your mini pyramid and it will be preserved; some "pyramidiots" even wore small pyramids on their heads to be energized.[9] The movie *Stargate* uses the Pyramid as a launchpad to the next world. The ancient Egyptians would have thought all this very funny. For them the pyramid shape had hardly any significance, certainly nothing magical. It was simply an architectural development, the same way our skyscrapers evolved out of smaller earlier buildings. For the Egyptians there was one and only one purpose for a pyramid—to protect the body of the pharaoh. It was all about life after death.

No civilization has ever devoted more of its resources and energy to preparing for immortality than the Egyptians. Much of what we know about life in ancient Egypt comes from studying their physical preparations for life in the next world. They were resurrectionists—they believed they were going to get up and go again in the next world, where the party would continue forever. Because the next world was going to be a continuation of this one, you would need pretty much the same stuff you had in this world—clothing, food, furniture, even your dog. In 1906, the great Italian Egyptologist Ernesto Schiaparelli discovered the intact tomb of the architect Kha and his wife Merit. There, neatly folded, were all the clothes the couple would need for their journey to the afterlife complete with patches sewn on by Kha's wife. In one corner of the tomb was the board game that Kha and Merit played in the evenings, and with it the stools they sat on. Because Kha was an architect, he couldn't think of going to the next world without the cubit stick he used to measure his building projects. It's all there in the Egyptian Museum in Torino, Italy, packed by Kha and Merit for the future.[10] They were literally going to take it with them.

But what good were all the clothes you had packed for eternity if you couldn't wear them? You needed your body. Enter mummification. As every sixth grader can tell you, skilled embalmers removed the brain through the nose with a long metal hook, the internal organs through a small abdominal incision, and then they dehydrated the body so it was preserved and could reanimate in the next world. We know most of this by examining mummies found in tombs. Like pyramid builders, the embalmers never committed the details of their craft to papyrus. I was able to fill in some additional details of the process in 1994 when I mummified a human cadaver in the ancient Egyptian manner.[11]

The embalmers weren't the only ones involved in the immortality business. There were miners to dig the salts used to dehydrate the bodies, tomb cutters, artists to decorate tomb walls, coffin makers, scribes to write books of the dead, and priests to recite the prayers and perform the rituals needed for resurrection. All this cost money, but the Egyptians had it. Egypt was primarily a nation of farmers living along the Nile, but there was also a large middle class that could afford preparations for the next life. Because Egypt had a strong central government (the pharaoh), there was organization and taxation. Farmers grew crops and an army of bureaucrats recorded information about the crops, collected taxes, oversaw shipments to government granaries, and made sure everything was running smoothly. Add to this the hierarchy of priests, high priests, temple overseers, and other religious professionals and you have a large middle class who can afford a nice tomb to house the possessions they are taking to the next world.

Death was big business in ancient Egypt, and its biggest manifestation is the Great Pyramid of Giza built for one, and only one, purpose—to house the body of the dead king. The pharaoh, the living Horus, King of Upper and Lower Egypt, needed a tomb that would protect his mummy and all the goods he would take with him to the next world, and so the pyramid was created. But it didn't happen the way most people think it did—a bright young architect waking up one morning with the idea of building a pyramid. Rather, the pyramid shape was the result of hundreds of years of architectural development. It evolved; it wasn't invented. To understand the Great Pyramid, you have to understand the evolution that led up to it, and the beginning of that evolution is, surprisingly, in London.

One of the British Museum's most popular attractions is a dead Egyptian nicknamed Ginger because of his light-colored hair. Ginger died more than 5,000 years ago, centuries before embalming was invented and the first mummies were created, but still, he is well preserved. If you had known him when he was alive, you could still recognize him today after all those centuries. Ginger is a natural mummy, the

"Ginger," a 5,000-year-old mummy in the British Museum, was preserved naturally by burial in the dry Egyptian sand.

result of his burial in the dry Egyptian sand. In prehistoric times bodies like Ginger were buried in sand pits in the desert. The sand dehydrated the bodies quickly, before they could be attacked by bacteria, preserving them as well as most artificial mummies produced later in the embalmers' workshops. Ginger lies next to some of his possessions—pots, a reed mat, a necklace—suggesting that even as early as Ginger's time, there was a belief in the next world. Just a few centuries after Ginger's modest burial, the Egyptians would be building pyramids. But it was not a giant leap, not an all-at-once breakthrough; it was a step-by-step journey from Ginger's burial to the Great Pyramid of Giza.

The problem with being buried in a sand pit is that the bodies don't stay buried—sand blows away, exposing the body to animals. Even today, if you walk off the tourist paths at Saqqara, ancient Egypt's largest cemetery, you will sometimes see human bones protruding from the sand. Consequently, the next advance in ancient Egyptian burial practices was to bury the dead not in sand, but in bedrock. Clear away the sand, cut a deep shaft into the bedrock, and dig a burial chamber beneath the ground. Once the body and all the grave goods were in the burial chamber, the shaft was filled in with rubble to protect the body and its possessions. Then, on top of the shaft, above ground, a

Because of their rectangular shape, early Egyptian tombs were called *mastabas*—Arabic for "bench."

chapel was erected where the family could visit the deceased, make offerings, and pay their respects. Because these chapels are rectangular, the modern Egyptians call them *mastabas*—Arabic for bench.

They don't look like much from the outside, kind of like warehouses built of limestone blocks, but inside they are spectacular. The "warehouse" is divided into rooms, each about the size of a large living room. Some mastabas have a dozen or so rooms, imitating what the house of a well-heeled nobleman of 4,500 years ago would have looked like. It's not the layout that is spectacular, it's what is on the walls. Beautiful carvings show the daily life of the ancient Egyptians, letting the gods know how to treat the deceased. If a man liked duck hunting in the marshes, there he is on the wall, standing in his papyrus skiff, flinging his throw stick at a duck. Scenes along the Nile are shown in such detail that you can identify the species of fish beneath the water. In one scene a hippopotamus is giving birth, with the newborn emerging from its mother. Eagerly awaiting the birth is a hungry crocodile.

Other walls show the deceased with his entire family—wife, sons, daughters, even his servants. It's the ancient Egyptian equivalent of a family photo. The idea was that if you showed everyone together, then they would all be together in the afterlife. The deceased wasn't buried in these wonderfully decorated rooms; he was in the chamber beneath the mastaba. In the aboveground rooms, the family could occasionally gather for a meal and pay their respects to the deceased. The Egyptians had a saying, "To say the name of the dead is to make him live again," and that's just what would happen in these rooms.

Mastabas play an important part on the path to pyramids. From Ginger's humble burial in sand to the chapels of the noblemen of Egypt's Old Kingdom, there is no great conceptual leap, merely an obvious progression. Even I could have come up with the idea of building a small structure on top of a burial to protect it. We can all understand why this was done. Jean-Pierre was wrestling with a step quite a bit farther down the road, a step that was, indeed, a quantum leap. To me his internal ramp theory seemed brilliant, but I'm not a pyramid expert. I knew that if this unknown architect from Paris was going to be taken seriously, he needed the support of a real pyramid expert.

Meeting with the Master

Dieter Arnold works on the first floor of New York's Metropolitan Museum of Art. To get to his office you walk through the Met's fabulous Egyptian galleries. Right in the middle of the collection is an unmarked door painted the same color as the wall. Behind this door is a world the public never sees: the offices of the Egyptian Department. It is a duplex maze of offices, all with books stacked high on desks, posters on walls advertising exhibitions of Egyptian art at museums around the world. This is where the curators plan exhibitions, research objects offered for sale to the Museum, and where Dieter Arnold hangs his hat when not excavating in Egypt.

Dieter literally wrote the book—*Building in Egypt*—on ancient Egyptian construction techniques, but he is not an ivory tower academic. Now in his early sixties, he looks much younger and still retains the muscular physique he developed climbing the Alps in his younger days. With his mane of black and silver hair, he is not what the public expects a brilliant scholar to look like. He has spent his career excavating and rebuilding pyramids and moving large blocks of stone. When I have a question on pyramids, I turn to him. So I called Dieter and explained that Jean-Pierre seemed like the real deal to me. The next day, Jean-Pierre, again dressed in black, met me at the Metropolitan Museum of Art and behind the unmarked door I introduced Jean-Pierre to Dieter.

Dieter was in the middle of writing up his season's excavation report and I knew his time was precious. After a season in the field moving stones, recording inscriptions, and restoring walls, the Met's excavation team take their notes and prepare them for publication. This is not an easy task and involves coordinating many people with different skills—photographers, artists, excavators, translators of hieroglyphs; all have to combine their results to form a coherent picture of the season's findings. This was the process Jean-Pierre and I interrupted on a sunny summer day in New York.

Dieter was warm and welcoming, and Jean-Pierre was quickly doing his thing on his laptop. I hung back to watch Dieter's reaction. At first he was quiet, once in a while interjecting a "Ja, ja" in his slight German accent, then asking a few questions about a concern he had about the theory, and with each of Jean-Pierre's answers, another "Ja." Soon Dieter was animatedly discussing the theory. I had asked for fifteen minutes, but we stayed for more than an hour. Dieter never said he agreed with the theory, but clearly he thought it was worth considering. Jean-Pierre had passed the test. As we were leaving, Dieter casually mentioned that at one of his early pyramid excavations in the 1980s they had found traces of something that looked like an internal ramp, but didn't know what to make of it. Jean Pierre's theory was looking more and more probable.

During our meeting it became clear that while Jean-Pierre probably knew the Great Pyramid better than any one in the world, his focus was so narrow that he hardly knew any Egyptology. He thought there must be papyri discussing specific aspects of pyramid building. Dieter and I explained there were none. Established pyramid experts had the big picture, understood the context in which the Pyramid was built, knew what the civilization was like, but Jean-Pierre had a bad case of tunnel vision. He knew the blocks of the Pyramid and the techniques needed to build the Pyramid better than any of the experts, but aside from the nuts-and-bolts details, he was a babe in the woods. Even today, after hundreds of hours with him, I am not sure Jean-Pierre is really interested in Egypt; the Pyramid is his passion. It was clear that he needed help to get his theory tested.

As Jean-Pierre and I parted on the steps of the Metropolitan Museum, he was elated. He knew that having someone as well regarded as Dieter on his side was crucial to his credibility. The meeting with Dieter convinced me that Jean-Pierre's theory could be the most important archaeological discovery of our time. My job would be to make connections for him, introduce him to other people necessary for the project's success, and help him navigate the very narrow channels of Egyptian bureaucracy.

The internal ramp theory was different from other Egyptological discoveries. Tutankhamen's undisturbed tomb is the gold standard for all archaeological finds. Think of all the treasures, the fabulous art, the gold mask and coffins. It doesn't get much better than that. But in a way, Jean-Pierre's discovery is even better. True, it will not produce any artifacts. If the ramp is indeed inside the Pyramid, it will almost certainly be empty. The excitement of the internal ramp is intellectual. If the theory is correct, it gives us a window into one of the greatest intellectual achievements of all time; it shows just how advanced ancient Egyptian architects were in planning the pyramids, how far ahead they had to visualize to overcome incredible obstacles. Indeed, if Jean-Pierre is right about the details of construction, the Pyramid might just be the most extraordinary engineering accomplishment of all time, a monument in stone to what the human mind at its best is capable of. When I first heard Jean-Pierre's theory I called some Egyptologist friends to try it out on them. They too were excited. They weren't all convinced, but clearly this was big news. Convinced that Jean-Pierre's theory was the most important project I could work on, I happily put my own research on the back burner. If his theory was correct, I wanted to be there when he took his long walk up that mile-long internal ramp.

Imhotep Builds the Step Pyramid

Saqqara, Egypt, 2668 B.C.

Our architect is Imhotep, but he was far more than just an architect. He was the royal physician who would later be deified by the Greeks and Romans as Aesculapius, the god of healing; he was the vizier, the equivalent of the prime minister of Egypt, the greatest nation on earth. He was the Leonardo da Vinci of Egypt, the first recorded genius in history. He also had the puzzling title of Maker of Stone Vessels—perhaps a hobby to unwind after a hard day at the office(s). Imhotep lived at the beginning of Egypt's recorded history, at the time when Egypt was just realizing its greatness. The pharaoh he served was King Zoser. (For a brief description of the search of Imhotep's tomb, see Appendix I.)

Egypt dominated the Near East for two reasons: the pharaoh and the Nile, the more obvious being the Nile. Each year the Nile overflowed its banks, depositing rich, dark topsoil, fertilizing the land. With this gift, Egypt could grow more crops than were needed to feed its population of about one million. With all this food, Egypt could afford a professional standing army. Other countries had to make up their armies from farmers, carpenters, and whoever else could be rounded up when it was time to fight. Egypt had its army ready and waiting;

no country could withstand such an army. So each year, the Egyptian army marched out, terrorized other countries, and returned home with anything that wasn't nailed down. War was a significant part of the Egyptian economy and the Nile made this possible. But the Egyptians had yet another advantage over neighboring countries: the pharaoh.

Ancient Egypt was originally divided into two segments, Upper and Lower Egypt. Then sometime around 3200 B.C., a king from the south named Narmer marched north, defeated the northern king, and unified Egypt into the first nation in history. The story of Narmer's conquest is told on Egypt's equivalent of the Magna Carta, the Narmer Palette. The palette has pride of place on the first floor of Cairo's Egyptian Museum, just past the entrance. Carved from a single piece of slate, the palette is about two feet high and three inches thick. On the first side, Narmer is shown wearing the white crown of the south as he defeats his northern counterpart. On the flip side, Narmer leads a triumphal parade and wears the red crown of the north. He is the first king of both Upper and Lower Egypt. Symbolizing this unification, the palette shows two mythological creatures with their long necks intertwined. Egypt is a single nation and Narmer is its king. For 3,000 years, the icon of Egypt would be the pharaoh smiting his enemy, just as it appeared on the Narmer Palette.

The Narmer Palette is the world's first historical document. On one side King Narmer wears the white crown of the south, showing he is king of that region. On the second side of the palette, Narmer has conquered the north and wears the red crown. Egypt has been unified.

From this point on, Egypt would have a king as its ruler, but not an ordinary king. Unlike other kings of the ancient Near East, Egyptian pharaohs were gods. Pharaohs had the title King of Upper and Lower Egypt, but they were also Son of Re—the sun god. Never had such power been concentrated in the hands of one man. Even the calendar revolved around the pharaoh. When a new pharaoh was crowned, the calendar began anew—day 1, year 1 of the reign of Sesostris. When Sesostris died, the calendar was reset: day 1, year 1 of the reign of Amenemhet.

So when Imhotep began building a tomb for his pharaoh, he was in a unique position. Not only was he the royal architect, but he was also the prime minister of the wealthiest nation on the planet. He had all the money he needed and didn't have to worry about the approval of committees. If the pharaoh wanted it, it would be built. The only limit was his imagination, and the brilliant Imhotep conceived of something totally new, a building in stone.

Like his ancestors, Zoser was going to be buried beneath the ground with a huge mastaba above the burial chamber, but his mastaba was not going to be mud brick; it would be stone—the first significant stone building in history. As the mastaba grew, Imhotep gained confidence in his ability to build in stone and increased its size. Then came his second big idea. He would place another mastaba on top of the first, and then another, and then three more, creating the Step Pyramid of Saqqara, the world's first pyramid. At 240 feet high, it is as tall as a modern twenty-five-story building, its base covering five acres. Probably ten times taller than any other structure in Egypt, it must have been a source of pride for the entire nation. Remember, this was a country where even the king lived in a mud brick palace. Then, all of a sudden, a huge stone building rises out of the desert. It must have been incredible.

As fantastic as it was, there are signs that the Egyptians were just learning how to build in stone. The Step Pyramid is solid, there are no chambers inside, and two clues suggest the Egyptians hadn't fully mastered stonework. First, the stones are not squared, their sides are not precise, and their corners aren't true right angles. Because they are irregular, they don't stack perfectly and some of the forces are directed outward rather than downward, making the pyramid unstable. To keep it from collapsing during the later stages of construction, Imhotep slanted the external walls of the pyramid inward to counterbalance

The Step Pyramid of Saqqara is the first stone building in history

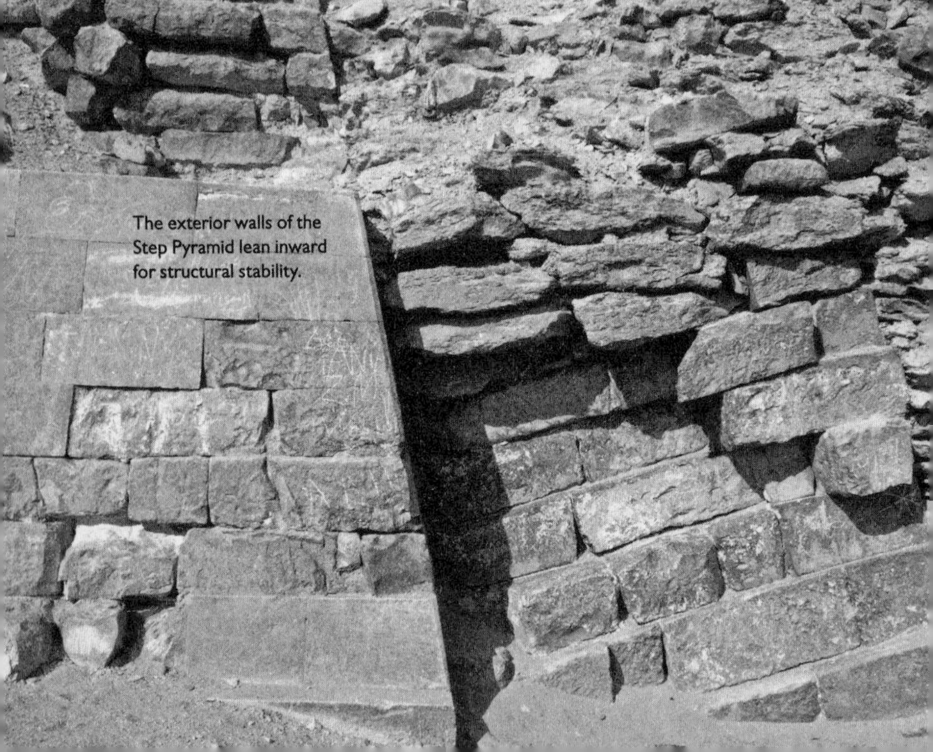

The exterior walls of the Step Pyramid lean inward for structural stability.

the outward forces. With all the stones angled toward the center of the pyramid, it is literally leaning in on itself.[12]

The second indication that Imhotep was just learning how to build in stone is the size of the blocks. They weigh less than 100 pounds each, small compared with the two-ton blocks that would be used in the Great Pyramid. Imhotep and his workers had not yet figured out how to move and lift massive stones; they were just feeling their way. The blocks look like giant mud bricks, only carved in stone. The Egyptians were imitating what they knew and were used to.

The Step Pyramid does not stand alone in the desert. Imhotep built a virtual city in stone for Zoser, complete with chapels, houses, statues—literally a petrified city. The pyramid complex is surrounded by a stone enclosure wall twenty feet high. Every few yards a false door was cut into the wall to look like a door, but there's only one real entrance. When you step through this entrance, there are signs all around that you are at the birthplace of building in stone. Most tourists don't see it; you have to know what to look for. You walk into a long hallway, but really you are inside a stone model of what an ancient palace built of wood and reeds looked like. Above your head, the ceiling sculpted in stone imitates the wooden beams of a palace. To the left and right of you are huge doors twelve feet high and eight feet wide, complete with hinges, but they will never swing shut; they are constructed of stone blocks. Think of giant Legos and you won't be very wrong. As you continue into Zoser's pyramid complex, you walk between two rows of stone columns, carved to look like bundles of papyri. Look closely and you will see that these columns are not freestanding, they are "engaged," attached to the walls; Imhotep probably wasn't sure if stone columns could stand alone.

The Step Pyramid is an incredibly bold venture for the first building in stone. You would expect smaller stone structures to come first, and then increase in size as Egypt worked up to something the size of the Step Pyramid. But no, there are no earlier stone buildings in Egypt. It's not that they're out there and archaeologists just haven't found them yet. There are plenty of remains of early buildings and tombs, but they were all mud brick. If there were stone buildings, some would have been preserved. With no prior experience, Imhotep constructed the first building in stone—and the largest structure in the world.

The pyramid is an impressive monument, but the burial chambers beneath it are just as impressive. The corridors and rooms stretch for more than three miles. Imagine one of those crazy English maze gardens, only much bigger and carved out of bedrock a hundred feet below ground, and you'll have some idea of where King Zoser was laid to rest. It's more like something you'd expect beneath the Pentagon than the burial site of an ancient king. It's mostly corridors, but every now and then there is a rectangular room cut from the rock to hold the pharaoh's treasures. Some of the rooms are still crammed with thousands of stone pots and vases for Zoser's journey to the next world. Inside the burial chamber, carved reliefs show the king running during the ritual of the Heb-Sed Festival, ensuring that he will be young forever. Some walls are lined with beautiful blue-green ceramic tiles, imitating the woven reed mats that covered the mud brick palace walls. Again, Imhotep was imitating the building materials of the living in a more durable substance suited for eternity.

Imhotep conceived many innovations for his pharaoh's burial place; the most puzzling is called the Southern Burial. A quarter of a mile south of the Step Pyramid is a second burial chamber for Zoser. It is not a pyramid, and today just looks like a deep hole in the ground—a shaft leads to a single burial chamber carved in the bedrock. Here too beautiful blue-green ceramic tiles decorate the walls, and there are scenes of Zoser running in the Heb-Sed Festival, but the chamber is really too small for a burial. Some have suggested that Zoser had two tombs, one northern, one southern, to emphasize that he was king of both Upper and Lower Egypt. The southern one was a cenotaph, a false burial chamber.

SNEFERU: KING OF THE PYRAMIDS

Meidum, Egypt, 2613 B.C.

Imhotep's great creation, the Step Pyramid of Saqqara, sparked a frenzy of construction in Egypt, and a nation of farmers became a nation of builders. It is even possible that Imhotep went on to design a second step pyramid for Zoser's successor, King Sekhem-Khet, whose pyramid was discovered in the 1950s (see Appendix II). After building several step pyramids, Egypt would go on to erect even greater monuments—and one of the strangest of these is the Meidum Pyramid.

The pyramid at Meidum sits isolated in the desert about fifty miles south of Saqqara. Rarely visited by tourists, it is a crucial step in Egypt's march toward the Great Pyramid. As soon as you see the Meidum Pyramid you know something went wrong. Looking more like a medieval fortress than a pyramid, it seems almost sinister. In the 1960s, Kurt Mendelssohn, an Oxford University physicist, theorized that the walls of the pyramid collapsed during construction because the angle of the pyramid was too steep.[13] He believed the mound of sand at its base hid the top of the pyramid that came crashing down. However, recent excavations of the mound show that the collapse theory is wrong; the mound consists primarily of windblown sand. Egyptologists now agree that the reason for the pyramid's ruinous state was that local villagers used it as a quarry for stone, stripping it of its fine white limestone casing. But there's still a mystery. The pyramid was never used for the pharaoh's burial, and no one knows why.

There is a temple next to the pyramid where priests would have made offerings for the dead king. On top of the temple are two stelae—round-topped stones that served as ancient Egypt's bulletin boards. If you wanted something known, you carved it on a stela and put it where everyone could see it. The two stelae at the Meidum Pyramid should have the king's name and titles, but they are totally blank; they were never inscribed—a dead giveaway that the king never used the pyramid. The unfinished burial chamber inside the pyramid offers another clue that the pyramid was never used, but within it rests a milestone in the history of pyramid building.

The burial chamber inside the Meidum Pyramid is the first aboveground burial in Egypt, a radical break from the underground burial chamber concept of the Step Pyramid. The owner of the Meidum Pyramid was going to be buried *in* the pyramid, not *under* it. To be buried inside a pyramid, a major engineering problem had to be solved. If the burial chamber is inside the pyramid, then the ceiling of the chamber must support the hundreds of thousands of tons above it. Constructing a room inside a pyramid had never been tried before and the architect of the Meidum Pyramid came up with an ingenious solution—a corbelled ceiling.

With a corbelled ceiling, the walls narrow as they get higher. As you build the wall out of stone blocks, each level is placed about six inches in from the one beneath it, so it overlaps and looks like an upside-down staircase. Thus, when you get to the top, the block spanning the walls and forming the ceiling is only a few inches wide. A block only a few inches wide is not going to crack under the weight above it, thus the problem of how to build an internal room is solved.

The innovation of a room inside the pyramid was not the only engineering advance at Meidum; the Meidum Pyramid is also the first attempt at a true pyramid. It began as a stepped pyramid, but some time near the end of its construction the architect had the idea to fill in the steps. We don't know why the project was abandoned, but there must have been a very good reason—a significant percentage of Egypt's economy had been poured into creating the largest building on earth,

A corbelled ceiling in the Meidum Pyramid made it possible to have a burial chamber *inside* the pyramid.

yet it was never used. Some think the outer casing blocks were not securely tied in to the rest of the pyramid and began slipping, but this is far from certain.[14] We do know that the pharaoh was never buried in the unfinished burial chamber, and by a stroke of good luck, we know who this pharaoh was.

During the 18th Dynasty—about a thousand years after the Meidum Pyramid was abandoned—a scribe named Akheperkare-seneb visited the abandoned pyramid. By this time, the era of pyramid building was long over and no doubt he walked around wondering at its construction. He would have strolled down to the Valley Temple, about a quarter of a mile from the pyramid. It was probably intended that the deceased pharaoh be mummified inside the temple, but it too was never used. Akheperkare-seneb walked up the causeway connecting the Valley Temple with the pyramid and came to the mortuary chapel next to the pyramid, where priests would have made the daily offerings that ensured continued existence for the soul of their king. He paused in this chapel as sunlight streamed in. He took his scribe's palette, dipped his reed brush into a bit of water, touched it to his block of black ink and wrote on the wall: "On the twelfth day of the fourth month of summer, in the 41st year of King Tuthmosis III, the scribe Akheperkare-seneb, son of Ammenmesu, came to see the beautiful temple of King Sneferu. He found it as though heaven were in it, and the sun rising in it." Then he added: "May heaven rain with fresh myrrh, may it drip with incense upon the roof of the temple of King Sneferu." Because of this ancient graffito, we know that the Meidum Pyramid was built by King Sneferu, but this was not his only pyramid.

Abandoning the Meidum Pyramid left Sneferu without a burial place, so a second pyramid had to be constructed. It is even possible that while the Meidum Pyramid was under construction, Sneferu's second pyramid was also being built. After all, Zoser had two burials, the Step Pyramid and his "southern burial," so Sneferu may have always intended to have two pyramids. The second one, at Dashur, about thirty miles north of Meidum, is the first pyramid designed from the beginning to be a true pyramid, one with smooth sides.

There is a popular tendency to credit the Egyptians with more knowledge than they actually had. Even the Greek philosopher Plato believed that the legendary mathematicians of his past such as Thales of Miletus and Pythagoras had studied in Egypt.[15] The truth is, the Egyptians only had very basic mathematical skills and probably didn't teach the Greeks anything.[16] It doesn't take higher mathematics to build a pyramid, but it does take manpower and almost superhuman precision, which we will see when we examine the construction of the Great Pyramid. Ancient architects had no way to calculate load-bearing capacities of various materials, so they worked by trial and error. Keep piling blocks on top of a granite beam until it cracks; then you know what it can support. This kind of construction can lead to disaster, and in the history of pyramid construction no disaster stands out more than Sneferu's second pyramid, the Bent Pyramid of Dashur.

Pyramids may all look alike, but each one is unique, with each architect trying to outdo the other. The Bent Pyramid is the first pyramid to have two entrances and two burial chambers—it's the new, improved, deluxe model. When they were built, the burial chambers were the most spectacular interior spaces on earth. The corbelling used so tentatively in Sneferu's pyramid at Meidum is perfected at Dashur, where the Bent Pyramid stands. All four sides of the rooms are stepped inward so that the entire room gradually narrows toward the top, yielding a dizzying view to the ceiling fifty-five feet above the floor. One of the two burial chambers was undoubtedly intended for Sneferu and the other was probably for his Great Wife, Queen Hetepheres, but neither was ever buried inside. Incredibly, this pyramid was also abandoned, but in this case we know why.

About two-thirds up the face of the pyramid the angle of incline bends, giving the pyramid its distinctive shape and name. The bend is the result of one of the costliest engineering disasters ever. Pyramids are

The corbelling inside the Bent Pyramid soars upward for fifty-five feet on all four sides. In the center is the ladder build by the Antiquities Service.

never constructed on sand; sand shifts, blocks move, and the pyramid would collapse. Pyramids are built on bedrock, and the Bent Pyramid is no exception. However, one corner of the pyramid rested on a layer of gravel, making that one corner unstable. During the early stages of construction there were no problems, but as the pyramid grew taller and its mass increased, the weight from the blocks on the corner resting on gravel caused those blocks to shift. This movement was transmitted inside the pyramid and the walls of the upper burial chamber cracked and started moving inward. In a desperate attempt to stop the room from imploding, the ancient engineers wedged huge fifty-foot cedar of Lebanon logs between opposite walls to keep them from collapsing. This stabilized the pyramid, but clearly it was no longer suitable for Sneferu's burial. It could, however, serve as his symbolic burial, one of the two burials that pharaohs now had, one for Upper and one for Lower Egypt. The pyramid was completed as quickly and inexpensively as possible, and this is why its angle was changed. Having a gentler slope at the top greatly reduced the number of blocks in the top third of the pyramid and also reduced the weight on the fragile burial chamber. I can just imagine the discussion between the architects when the burial chamber cracked: "You tell the pharaoh." "No, *you* tell the pharaoh." Sneferu was now in a very difficult position. He had built the two largest buildings on earth and neither was suitable for his burial. Time was running out for the aging pharaoh.

Sneferu did not give up on pyramids. With two disasters behind him, he built his third pyramid at Dashur, just a mile from the unused Bent Pyramid. Because it gleams red in the sunlight, it is known as the Red Pyramid. There are clear indications that Sneferu wasn't taking any chances with this one; it had to succeed. Unlike all the earlier pyramids, it has a gentle 42-degree slope compared with 53 for the Bent and Meidum Pyramids. The gentle slope accomplished two things: first, it reduced the number of stones needed to build the pyramid. Sneferu was getting old and wanted to reduce building time. Second, a gentler slope reduced the chances of more construction disasters. In the Red Pyramid, there are no indications of construction problems, and Sneferu was undoubtedly laid to rest inside it, the first true pyramid in history.

Sneferu's three pyramids—Meidum, the Bent, and the Red—constitute the most intense building spurt in history. Under one pharaoh,

Egypt built the three largest buildings on earth, developed the first true pyramid, and invented the corbelled ceiling so burials could be above ground, high up in the pyramid. In television programs and popular books about pyramids, the focus is almost always on King Khufu, the builder of the Great Pyramid at Giza. However, it is really Khufu's father, Sneferu, who taught Egypt how to build pyramids. A thousand years after Sneferu's death, when someone did something impressive, he would say, "Not since the time of Sneferu has its like been done."

In ancient Egypt, the profession of architect was often handed down from father to son. The architects of the largest pyramids ever built—the Great Pyramid and the pyramid of King Kephren—were uncle and nephew. When boys see their fathers and uncles building great things, they want to also. So it was with Jean-Pierre and his father.

The Red Pyramid was the third pyramid built by King Sneferu.

An Architect Is Born

In the years immediately following World War I, France experienced a strange but understandable birth pattern. Families were having only one child. The country had been devastated by the war, times were hard, and everyone's energies were directed toward rebuilding the country. The phenomenon affected French society for generations. With the typical nuclear family of father, mother, and only child, much of the following generation would grow up with no uncles, aunts, or cousins. These "only children" born after World War I would come of age during World War II and bear much of the brunt of that war. Jean-Pierre's father was one of them.

Rushed through engineering school so he could help the war effort by building runways for the air force, Henri Houdin gained a lot of experience very fast. The same year he was graduated from the École des Arts et Métiers, Renée Mesana was graduating as a physician and setting up her practice near Paris. With many male physicians away at war, her services were urgently needed for the civilian population and she too was gaining experience rapidly. One of her first house calls just after the war was to treat Henri's sick grandmother, but Henri and Renée never spoke. He did, however, notice "the pretty young physician in the convertible."

When the war finally ended there was rejoicing, but much of France had again been destroyed by German bombs and had to be rebuilt. If you could build a bridge or repair a road, you had a promising future. With more than 7,000 bridges in need of rebuilding, young engineers were given tremendous responsibilities. Thus, in 1947, twenty-four-year-old Henri Houdin was placed in charge of rebuilding the Conflans Bridge outside of Paris. The first of his many successes, the bridge was completed on schedule and within budget. Soon he was being given other major projects and responsibilities. Although they hadn't yet spoken, Renée's career was paralleling Henri's; her practice was growing and her skills increasing at a postwar pace. When they met at the wedding of a mutual friend, the two young professionals were immediately attracted to each other; in less than a year they married. As they settled into their Paris apartment and planned their family, neither could have foreseen that their children would be raised in Africa.

In 1949 Bernard, their first child, was born, and the following year Henri was sent to Ivory Coast to evaluate building projects. Ivory Coast was a French protectorate and with the war behind her, France had the resources to develop her colonies. Henri discovered that Ivory Coast had such enormous infrastructure needs that an entire company had to be established for all the building. Soon, Henri, Renée, and baby Bernard were settled in Abidjan.

In 1951 Jean-Pierre was born, and now the family was complete. When she wasn't taking care of the two boys, Renée served as company physician, treating sick or injured workers and anyone else needing a doctor's services. Henri was constantly building—roads, dams, water towers, schools—but because of his experience in France, bridges became his specialty. France still had plenty of war surplus and several of the bridges in Ivory Coast were built from Mulberry bridges constructed for the D-Day invasion at Normandy. Henri's biggest project was the huge Houphouet Boigny Bridge linking two districts of the capital, Abidjan, by spanning a lagoon.

Before the bridge could be built, a study had to be completed to make sure that the underwater soil on which the piers would rest could support a bridge. The government sent Jean Kerisel, a ground specialist, to perform the necessary tests. Soon after his arrival Kerisel fell ill and was eager to return to France, but Jean-Pierre's physician mother cured him and he remained in Ivory Coast to complete his study. A bond

Henri and Renée Houdin in Africa in the early 1950s.

formed between engineer Henri and the ground specialist. Perhaps it was the fact that Henri's wife had cured Kerisel, perhaps it was the shared difficulties of building the bridge, but the two men remained friends for life. Nearly half a century after their first project together, both were trying to figure out how the Great Pyramid was built.

The bridge they built was an engineering marvel of its day, with a highway on top and two concrete tunnels acting as beams beneath the highway through which trains ran. Sections of the concrete tunnels were poured on the ground and then lifted into place by huge cranes on barges. During the three years it took to build, Bernard and Jean-Pierre frequently accompanied their father to watch the construction. When it was inaugurated in March of 1958 fabulous fireworks lit up the night sky, something the brothers would remember for the rest of their lives.

Bernard enjoyed visiting work sites with his father, but it was Jean-Pierre who wanted to know how everything worked. He grew up playing among bulldozers, cranes, Dumpsters, and trucks, all painted bright orange because in Ivory Coast the soil is reddish orange laterite and the dust won't show on orange-colored machinery.

Growing up in Ivory Coast in the 1950s was heaven for the two boys. They attended a neighborhood school and easily made friends with both the local boys and the children of the construction company's employees. There were soccer, swimming pools, and on weekends Henri would often fill the big Chevrolet Bel Air station wagon with the family and the boys' friends and drive to the beach or to a construction site for a picnic. Even illness could be turned into an adventure. An epidemic of whooping cough hit Ivory Coast, and both boys and two of the neighbor's children came down with it. In the 1950s a cutting-edge cure for the pulmonary complications of whooping cough involved decompressing the lungs on a high-altitude flight. Jean-Pierre's physician mother bundled the four children up and took them on a round-trip ride on a Douglas DC–3. All the patients survived. It was a wonderful, unforgettable childhood, but time was approaching for the brothers' higher education and Africa had to be left behind.

The Houdins returned to Paris and bought an apartment near the Arc de Triomphe, and both boys attended the Lycée Carnot, one of the most prestigious schools in France. Jean-Pierre struggled to get his baccalaureate, starting in mathematics but ending with a degree in philosophy. While the boys were in school, Henri was flying all over the world building bridges and roads in Ivory Coast, Tahiti, New Caledonia, Lebanon, and Greece. Renée returned to school to study industrial medicine. In the 1960s, way ahead of her time, she was one of the first experts to alert the public to the dangers of asbestos.

In 1970, Jean-Pierre entered the École des Beaux-Arts to begin six years of architectural studies. Finally life started in earnest. Working part-time in an architectural firm drawing plans and doing calculations, he learned the practical aspects of building while studying architectural design at school. After an undistinguished career at the Lycée, Beaux-Arts was a piece of cake. Every architecture student had to design a project in his last year and Jean-Pierre's avant-garde solar house was a hit with his professors.

Soon after graduation, Jean-Pierre took his plans for the solar house and boarded a DC–10 for Ivory Coast, intending to convince the government to go solar. They would have none of it, so he returned home and joined his father's construction business as an architect in residence. In 1987 the housing market declined, and Henri decided to close the company and retire. Jean-Pierre opened a private practice and continued to design condos and offices for clients, gaining experience in all aspects of construction. In the meantime Jean-Pierre also gained a wife, Michelle, a beautiful artist deeply involved in the Paris art community.

Jean-Pierre was doing well enough in his new architectural practice that he, Michelle, and an engineer friend named Laurent bought the bakery across the street from their apartment, and in 1986 opened a café called Les Enfants Gâtés (The Rotten Kids), that Jean-Pierre and Laurent designed as a 1920s salon. It was a great success and soon became a meeting place for artists, actors, and writers. The three introduced the idea of Sunday brunch to Paris and soon Americans were gathering there on weekends. The café was such a success that Michelle was able to open Gallery 43 in the limestone basement beneath the café. Young artists were invited to show their art free, with a preview offered in the café above. Les Enfants Gâtés launched many young artists' careers.

For nearly a decade, the café and gallery were a great success, but after ten years it was becoming old hat. Michelle feared they were "doing the routine." They had to move on, but to where and to what? Jean-Pierre's architectural practice had been good to them and the café was doing so well financially that they were able to sell it for enough money to take time off and search for something new. In the fall of 1996, the couple rented their apartment and left for New York in search of an idea. They were secure enough that they made their escape in style, aboard the Concorde. After hors d'oeuvres and champagne at the Concorde departure lounge with fellow passenger Calvin Klein and other celebrities, the adventurers boarded the plane at 11:00 A.M. and took off with a big push forcing them back in their seats like jet fighter pilots. Soon after leaving the coast of Brittany behind, the Concorde passed the speed of sound and kept climbing. After an hour the Machmeter in the front of the plane indicated Mach 2, twice the speed of sound, and the plane leveled off at 54,000 feet. Everything was silent and from the Concorde's small windows they could detect the earth's curvature. After a fabulous lunch, an announcement came through the cabin. They were beginning their descent into JFK and would be landing in twenty minutes, at 8:30 A.M. local time—two and a half hours earlier than they had left Paris. They had traveled faster than the sun.

Architect Adrift

New York, 1996

When Jean-Pierre and Michelle landed at Kennedy Airport they had no agenda, but enough money to hold out for a year. Comfortable in the art scene, they rented a friend's apartment in Greenwich Village. At first Jean-Pierre spent his days walking the city, taking in the architecture and energy. In the mid–1990s, New York was where things were happening, and although Jean-Pierre didn't know what was in store for him, he sensed he would find it here. On one of his wanderings he found himself looking into a window of a large computer store on Fifth Avenue. Fascinated, he went inside to look and left with a heavy computer in his arms.

The Internet was growing fast and Jean-Pierre realized this world might be for him—a computer, the Internet, no office, perhaps he could work anywhere. Enough of concrete and steel buildings, the computer was his road to a new freedom. At the same time that the Internet was growing, architecture was changing. Drawings and plans were no longer produced by hand. Now everything was being designed on computers. Fascinated with the new graphics programs available to architects, Jean-Pierre spent month after month learning, experimenting, seeing what could be done, and wondering how he could use these new skills.

He could design Web sites. While Jean-Pierre improved his computer skills, Michelle visited an average of twenty art galleries a week, taking in the New York art scene. She too was looking for a new direction and was interested in seeing how the new technologies were affecting art.

After several months in New York, they were ready to see America. They drove a rented car to Washington, D.C., to see the nation's capital, and then flew to Las Vegas. From there they drove across the American west—Arizona, Colorado, New Mexico—and it was there that Jean-Pierre became the Man in Black. It wasn't a fashion statement; it was a laundering decision. Michelle always wore black. As they traveled across the country they used Laundromats—it was easier if the wash was all one color.

In Colorado they visited Hoover Dam, one of the most massive construction projects in modern times. A more than 600-foot-thick base supports 700-foot walls. Built during the Depression, 3.3 million cubic meters of concrete—a mass greater than the Great Pyramid—were poured by 5,000 workers in less than five years. A modern engineering marvel, it had many similarities to the Great Pyramid, but Jean-Pierre wouldn't know that for several years. They visited San Francisco and Los Angeles, and then they were finally ready to return home. Almost exactly one year after they had departed on the Concorde, they flew back to Paris.

In October the tenant who had rented their apartment moved out, so they moved back into their wonderful apartment, a duplex in a seventeenth-century landmarked building in the Marais district of Paris. They were back in familiar surroundings, but not in the old routine. The American adventure had depleted their bank account, so it was time to go back to work. With his new computer graphics skills, Jean-Pierre began designing Web sites for architects with his engineer friend Laurent, who was also reinventing himself. Business was good; everyone wanted a Web site. Jean-Pierre had his freedom; he could go anywhere with his computer and earn a living.

By mid–1998 he and Michelle decided to cut the cord even further. They sold their apartment and began a life on the road, from apartment to apartment, no permanent address, no fixed duties. They would "live full time." The urgency for life came from Michelle. Both her parents had died early, and on the day after her mother's death in 1992, she turned to Jean-Pierre and said, "I would like to drink a little Champagne every evening." It is a ritual they have kept since that day.

Life was good. The idea Jean-Pierre was looking for had still not come to him. When it did, it would be thanks to two builders: his father and an ancient Egyptian architect.

A Troubled Bridge?

Abidjan, Ivory Coast, 1997

After Jean-Pierre's failed attempt to sell solar houses to the government of Ivory Coast, he never returned. His brother Bernard, however, maintained connections, and has dual Ivory Coast and French citizenship. On a trip to Abidjan he heard rumors that the bridge his father built was in danger of collapsing—the piers were weakening. He quickly informed his father and Henri notified the French authorities, as France was still responsible for the bridge. A committee to study the problem was established. Jean Kerisel, the engineer cured by Jean-Pierre's mother, was responsible for the piers and was placed on the committee, as was Henri Houdin. After forty years, the two were working together again. Their meetings led President Chirac to allocate funds to inspect and monitor the bridge. During this time in 1997, Kerisel told Henri of his new interest.

In the 1980s, Cairo was building its new underground metro system and Kerisel was on the team doing the soil survey. While in Cairo he made several trips to Giza to visit the Great Pyramid and became fascinated with the engineering problems that had to be solved to build such a monument. By now he was somewhat of an expert on pyramid building. In late 1998, Jean Kerisel was interviewed for a television documentary about pyramids and told Henri Houdin to watch for it.

The show, *The Mystery of the Pyramids*, presented by François de Closets, a well-known French television personality, was quite good. It covered the usual ground, discussing the possibility of the straight ramp and its problems and then the idea of a ramp corkscrewing up the outside of the Pyramid. Henri was intrigued and listened attentively as Egyptologists presented their theories of how the Great Pyramid was built. Ninety-year-old Jean-François Lauer, the elder statesman of pyramid experts, told his theory; right next to Lauer sat Henri's old friend Jean Kerisel. But Henri knew their theories were wrong. They were approaching the solution from the wrong side—the outside! Everyone seemed to think the pyramid is basically a huge sugar cube–type construction. Figure out a way to pile blocks up higher and higher and then just leave some out to form the chambers inside. But Henri's engineer's brain was kicking in; he knew it didn't work that way. You had to think about the interior first. Visualize the building of the interior chambers—the King's Chamber, the Queen's Chamber, the Grand Gallery. What construction techniques would be needed? What kinds of stone? How big would the stones be? Answers to questions like these determine how to build the outside of the pyramid. Excited by his idea, he called his architect son and blurted out, "The pyramids were not built from the outside, but from the inside."

Hemienu Plans the Great Pyramid

Giza, Egypt, 2589 B.C. (Year 1 in the reign of Khufu)

It was a momentous day when Sneferu, king of Upper and Lower Egypt, the Great God, died sometime around 2590 B.C., but the date of his death and the details of the funeral of one of Egypt's greatest pharaohs went unrecorded. This seems incredible to us, living in a society that wants every possible detail of Princess Diana's death and is still debating how JFK died, but things were different in ancient Egypt. Death was a defeat and Egyptian scribes only recorded victories. History was not intended to record objective facts; it was to present the glories of a nation for others to see and wonder at. This leaves Egyptologists with the task of sifting through fragments of information to piece together the details of a pharaoh's death.

For the reign of Sneferu, one of the most important fragments is the Palermo Stone, a chunk of black diorite in the Regional Museum of Archeology in Sicily. Originally the stone was more than six feet long, and inscribed on its polished surface were the names and reigns of more than 200 kings of Egypt. The fragment in Palermo is only thirteen inches wide and ten inches long, but it lists the earliest pharaohs, including Sneferu. It recounts the major events during the various kings' reigns and from it we learn that Sneferu sent a trading expedition to

Lebanon to obtain cedar for building boats and the doors of the great temples of Egypt. It must have been a successful mission; forty ships laden with huge logs returned home to Egypt. Fragments like the Palermo Stone are the bits and pieces from which Egyptologists reconstruct ancient lives, but they don't give us the exact date of a pharaoh's death. There are two reasons for this. As we noted before, the Egyptians viewed death as a defeat, but they also had a unique calendar.

The Egyptians didn't number their years consecutively. Our year 2008 will be followed by 2009, but in ancient Egypt when a new king like Sneferu ascended the throne, the calendar began anew with: "Day 1, Year 1 in the reign of Sneferu." The only reason we know Sneferu died in 2590 B.C. is that a few events such as total solar eclipses mentioned in ancient Egyptian records can be dated accurately in terms of our calendar. Let's say that an ancient Egyptian papyrus written during the reign of Ramses the Great mentions that a solar eclipse took place. Using our calendar, astronomers calculate exactly when the solar eclipse took place and then Egyptologists count backward to get dates for reigns of the earlier kings. Based on evidence like this, our best bet for Sneferu's death is 2590 B.C.

We can be sure, however, that when Sneferu died, all of Egypt mourned. Under his rule Egypt became an international power, sending trading expeditions to Lebanon for cedar and to the Sinai for turquoise and copper. Sneferu ushered in the era of the great pyramids, but there is another reason to believe Egypt mourned his passing. It is recorded on the Westcar Papyrus, located in the Egyptian Museum in Berlin.

Before the Egyptians invented papyrus, writing was done on clay tablets. After being inscribed, the damp clay tablets were baked in a kiln to be preserved. It was an expensive and tedious process to form a tablet out of clay, inscribe it with a stylus, and then bake it. Sending letters abroad was not easy; great care had to be taken that the tablets didn't crumble and break. The Mesopotamians even had special envelopes, also baked, to protect them. The invention of papyrus (from which we get our word "paper") created a literary boom. All of a sudden, writing was easy. Sheets of paper made from strips of the papyrus plant were glued together in long rolls that could be written on with a brush. No more baking of tablets, no problem transporting the writings—the publishing industry took off. The Egyptians wrote everything on papyrus, religious texts, battle accounts, magical spells, even fiction. The Westcar Papyrus, named after its owner, contains a series of magical stories told by Sneferu's grandson, Prince Bauefre, the Stephen King of ancient Egypt.[17]

One of the stories tells us that Sneferu was walking through the palace one day, feeling bored, with a "sickness of the heart." He called his palace magician, Djadja-em-ankh, and asked what he should do. The magician suggested His Majesty take a boat out on his pleasure lake and that he invite the beautiful young ladies of the palace to do the rowing. Sneferu liked the idea and improved on it. He commanded that twenty fishnet dresses be brought for the young ladies to wear as they rowed. The plan worked. "The heart of his majesty was happy at the sight of their rowing." However, suddenly everything stopped. One of the lovely rowers was upset, having dropped overboard a turquoise fish amulet that she was wearing. Sneferu offered to replace it; he wanted the rowing to continue, but the young maiden cried that she wanted only her lost amulet, not another.

Sneferu, once again, summoned the ingenious Djadja-em-ankh. The magician caused the waters of the lake to part, retrieved the amulet, and returned it to the young lady, and the rowing continued. Pretty impressive stuff from a palace magician! Remember, this is all happening a thousand years before Moses parted the Red Sea. But even as ancient Egyptian magical tales go, there is something extremely unusual about the story. It gives us insight into the personality of a pharaoh. Think about it. Sneferu is king of Egypt, a god on earth, but a palace handmaiden is comfortable enough in his presence to refuse his offer to replace the amulet. And to Sneferu's credit he doesn't respond with "Off with her head." He's concerned, understanding. He's a pharaoh we are supposed to like. It's hard for us to grasp how long ago Sneferu lived, but think about it this way: this story gives us the earliest anecdote ever about an historical character. Sneferu is the first individual in history.

Soon after Sneferu's death, his body was ferried across the Nile, from the east bank to the west. The west was associated with death because the Egyptian religion was primarily a solar cult and they believed that the sun god, Re, died in the west at the end of every day and was reborn at dawn in the east. Cemeteries were on the west bank, and this is where embalmers set up their workshops, far from the living. Sneferu's body was ferried across the Nile to the royal embalmers in an elegant funerary boat specially constructed for the king's last voyage. This huge 150-foot ship had a special cabin on the deck to shelter the king's body. We know what Sneferu's funerary boat looked like because of the sharp eyes of an Egyptian Antiquities Service official.

In 1954 the Egyptian Antiquities Service was clearing debris near the Great Pyramid of Khufu on the Giza Plateau when a young archaeologist, Kamal el-Mallakh, noticed that the ancient wall enclosing the Great Pyramid was not exactly symmetrical—the wall on the south side was fifteen feet closer to the base of the Pyramid than the walls on the other sides. Careful inspection revealed that the south wall had been built asymmetrically to conceal something beneath it. The wall was taken down, revealing forty-one massive limestone blocks covering a pit cut into the bedrock.

After the limestone blocks were removed, the excavators discovered a 144-foot-long boat that had been disassembled and carefully placed in the pit. The 1,224 pieces of wood had been stacked in thirteen layers and, amazingly, the wood was so well preserved that an attempt could be made to reconstruct the boat. Although it resembled a giant boat model kit, the pieces did not come with an instruction booklet. It took twenty years of trial and error to reconstruct it. In the process, a great deal was learned about how wooden boats were built in ancient Egypt.[18]

The Boat Beneath the Pyramid, as it became known, wasn't constructed like a modern boat. Modern boats are built from the inside out. A framing of ribs is nailed together and then the boat's hull—the outside planking—is nailed to the ribs.[19] Ancient Egyptian boat builders did the opposite. They built a boat from the outside in and never used ribs! The hull of Khufu's boat, found next to the Great Pyramid, is made of huge cedar planks, some seventy feet long and six inches thick. No trees growing in Lebanon today are tall enough to provide such boards. These planks were carved to create the curve of the boat. (Modern boat builders steam and bend the planks to get a curve.) You can still see the marks on the planks made by the boat builders' copper chisels as they shaped the boards. The planks were not nailed together as in later boat building, but were tied together with strong rope made from hemp. Once the hull was formed by these tied planks, wood struts

The Cheops boat may have been intended to ferry the dead king from the realm of the living to the realm of the dead.

were fitted inside the hull to give the boat structural strength. This kind of boat, held together entirely by rope, is called a sewn boat.

The reassembly led to another mystery. No one could figure out how the boat had been used. It didn't have a mast or sails, so it couldn't have sailed. Ten giant oars, fifteen feet long, were found with the boat, but there were no oarlocks, nor was it clear where the rowers could have stood on the deck. How was the boat propelled if it didn't sail and wasn't rowed? Another puzzle was the cabin on the boat's deck. It is totally enclosed, with no windows for ventilation or light. This dark and airless cabin provides a clue to the boat's actual function.

Let's start with the problem of how the boat moved. If it wasn't self-propelled, it could have been towed by other boats. We know from tomb paintings that some ceremonial boats were pulled by other boats that had sails and usable oars. But for what ceremony was Khufu's boat used? This is where the airless cabin comes in. The Khufu boat was used only once, for the pharaoh's funeral. We can easily imagine the body of the pharaoh inside the dark, airless cabin, shielded from the mourners who lined the banks of the Nile. This boat was found next to Khufu's pyramid and was his, but Sneferu would undoubtedly have had a boat similar to the Giza boat to transport his body to the embalmers' workshop in the west.

As Egypt mourned for Sneferu, a skilled team of embalmers worked in seclusion for seventy days to prepare his body for its final journey to the next world. They began by making a three-and-a-half-inch incision on the left side of the abdomen through which they would remove the internal organs. These professionals knew anatomy. They understood that the internal organs contained water that would cause the pharaoh's body to decay; in order for the pharaoh's body to last for eternity, all moisture had to be removed. Beginning with the spleen, the embalmers worked carefully, removing the intestines, stomach, and liver. Working upward, they cut through the diaphragm, entering the thoracic cavity, carefully removing the lungs without disturbing the heart. Sneferu would need his heart to remember the magical spells he had to recite in order to be resurrected. The Egyptians believed you thought with your heart, which was perfectly reasonable. When you get excited, it's your heart that beats quickly, not your brain. This is why on Valentine's Day we send little chocolate hearts, though we should be sending little chocolate brains.

Once the internal organs were removed, the embalmers began the most difficult of all the surgical procedures, the removal of the brain. They pushed a copper hook resembling a straightened metal coat hanger up through the nasal passage. Breaking a thin bone behind the eyes, the instrument entered the cranial cavity. The copper hook was rotated repeatedly until the brain was liquefied. The pharaoh's body was then held with the head downward and feet toward the ceiling so the brain could run out through the nose. With the surgical procedures completed, the embalmers moved to the next stage of mummification—dehydration of the body.

To remove the remaining bodily fluids, the embalmers covered Sneferu's body with 400 pounds of natron, a naturally occurring mixture of baking soda and salt. After thirty-five days, the body was fully dehydrated by the white powder that preserved it, and Sneferu was wrapped in hundreds of yards of pure white linen to protect his body. As the ancient morticians performed their tasks, priests chanted hymns to ensure Sneferu's resurrection and entry into the next world. After seventy days, the period prescribed by the Egyptian religion, Sneferu's mummified body was placed in his pyramid at Dashur.

By the time Sneferu's body was being ferried across the Nile to the west, the Egyptian calendar had already been reset to Day 1, Year 1 of his son Khufu's reign. And while the embalmers were mummifying Sneferu's body, a spot for his son's pyramid was being selected. We know the man who was in charge of the selection—Hemienu, Khufu's architect.

We know quite a lot about Hemienu because the walls of his huge tomb on the Giza Plateau are inscribed with his biography. Hemienu was Sneferu's son—"The king's son of his own body" and thus Khufu's brother, or at least his half-brother, as Sneferu had several wives. It is not surprising that Hemienu became an architect. As a young prince growing up in the royal household, he probably visited his father's pyramids as they were being constructed at Meidum and Dashur. He also would have seen the tremendous power wielded by his father's architect.

We call Hemienu an architect but this is not quite accurate. "Architect" is a Roman word that means arch builder. Arches were so central to Roman building that they defined the profession. In the ancient world, there was no clear distinction between architects, the people who designed buildings, and engineers, the people concerned with the technical aspects of those buildings. Hemienu's actual title was Overseer of the King's Works. That meant he was responsible for design but also oversaw the construction of all pyramids, temples, and palaces throughout Egypt. He, of course, had a cadre of other overseers of works under him and under them were master stonemasons, road builders, and transport experts. Not only do we know Hemienu's titles and responsibilities, but we know what he looked like. A larger-than-life-size statue in Germany's Hildesheim Museum shows Hemienu in middle age, bald and with rolls of fat around his waist, a sign of prosperity. It is not an accident of history that Hemienu's name has been preserved for us; architects were important people in ancient Egypt.

Hemienu, architect of the Great Pyramid, was shown middle-aged, with rolls of fat—a sign of prosperity.

In 332 B.C., at the end of her 3,000-year history, Egypt was conquered by Alexander the Great, and for the next three centuries was ruled by Greek pharaohs. The last Greek to rule Egypt was the famous Cleopatra VII. During this Greek period of domination, Manetho, an Egyptian priest who spoke Greek as well as his native tongue, decided to write a history of Egypt, in Greek, so that the rulers would know something about Egypt's marvelous past. He called his history *Agyptiaca*—Egypt—and it is centered round the deeds of the pharaohs.[20] Manetho was the first to organize the Egyptian kings into dynasties, a division still used by Egyptologists today. By the time Manetho was writing his history,

Hemienu had been dead for more than 2,000 years, but his achievements and those of other architects were not forgotten. In the rare instances when Manetho mentions a nonroyal name, it is almost always an architect.

As architect of the Great Pyramid, Hemienu's first task would have been to select the building site. Religion required that it be on the west bank of the Nile. Osiris, God of the Dead, was Lord of the West and when anyone died he was called a westerner. Hemienu knew that his brother's pyramid must be in the west, where all the other pyramids were. Religion was one consideration when selecting the site; politics was another. Memphis, the ancient capital of Egypt, was in the north, so it is logical that the Pyramid would be in the north, near the seat of power. Memphis was situated just south of Egypt's delta, a lush marshland teeming with fish and fowl that fanned out all the way to the Mediterranean. A major source of Egypt's protein, the delta was also a favorite hunting spot for the elite.

For many foreigners, Memphis *was* Egypt—there was no need to venture south. They would enter Egypt through the delta and unload trade goods—timber, oil, wine, copper ingots—at the docks in Memphis. The harbor was constantly bustling with shiploads of cedar logs, exotic dyes, and Mediterranean wares of all kinds. But Memphis wasn't just an economic hub; it was the seat of government and a thriving religious center. The pharaoh had a palace at Memphis. Administrative buildings of mud brick with hundreds of small rooms for scribes and officials stretched out along the Nile. Here men recorded the amount of grain produced throughout Egypt, taxes, and the pharaoh's generous donations to the temples. This was the place where every bright Egyptian wanted to work. If you were a student in a village school in the south, "making it" meant getting a job as a local scribe, being noticed by a visiting official, and moving north to Memphis.

Along with the palaces and official buildings were temples to the various gods. This was the capital of Egypt, so all interests had to be represented, but Memphis also had its own triad of patron gods: Ptah the creator god, Sekhmet his wife, and Nefertum their son. Each god had its own large temple alongside smaller temples for Egypt's other gods. It was a glorious gleaming-white city.[21] Memphis's earliest name, The White Wall, referred to the limestone wall that surrounded the entire city.

The largest city by far in all of Egypt, Memphis had another distinction. It was situated on the west bank of the Nile. Almost every large city was on the east, with the west bank being reserved for cemeteries. Perhaps Memphis was so old that it was founded before this custom was established. We can't be sure, but being on the west bank put this great city in close proximity to Saqqara—the largest cemetery in Egypt.

From the earliest times, Saqqara was Egypt's elite burial ground. Pharaohs of Dynasty 1 have tombs here and for thousands of years kings and royalty continued to be buried there. Saqqara is probably the densest archaeological site in the world. Dig anywhere beneath its hundreds of acres and you will find an important tomb. But this is not where Khufu was going to be buried. Religion and politics dictated the general area of the Pyramid; construction concerns determined the exact site.

Two million limestone blocks averaging two and a half tons each were used to build the Great Pyramid. Every mile between the quarry and the Pyramid would mean extra hours transporting each block, adding years to the construction time, so great care had to be taken when selecting the site. As soon as a pharaoh died, the next king began planning his own tomb. So sometime during 2589 B.C., soon after their father died, the architect Hemienu sailed up and down the Nile in the area of Memphis looking for a suitable spot for his brother's Pyramid on the west bank of the river where limestone was readily available. The ship that carried Hemienu would have been a large stately affair, similar in size and construction technique to the funerary barge found buried on the Giza Plateau. Hemienu's had a mast and sails for going south with the wind. For the northward portion, going with the current, rowers manned the oars and guided the ship.

Hemienu traveled with his team of skilled builders. An Overseer of the Quarries made sure there was plenty of suitable stone in the area. An Overseer of Transport determined if a harbor could be built on the site and if canals to the Nile could be easily dug. It must have been incredibly exciting as they all stood on the deck of Hemienu's boat discussing the project, weighing the merits of one site over another. They were about to begin the largest construction project in the history of the world, and it was all for the glory of their pharaoh. Building Khufu's Pyramid would consume all their energy and attention for decades.

A stroke of engineering genius led Hemienu and his team to select the

Giza Plateau. He would not build the Pyramid *near* the limestone quarry; he would build it *in* the quarry. In the time of Hemienu, the Giza Plateau was a vast, deserted, gleaming-white outcropping of limestone covering more than 100 acres. On the north side it dropped off precipitously to the sand below, but on the south side was a gentle slope leading to a second, even larger, outcropping of limestone more than three-quarters of a mile long and one-quarter-mile wide that could be used as a second quarry. By placing Khufu's Pyramid on the Giza Plateau, Hemienu could quarry almost all the stone he needed right on-site.

As Hemienu and his team stood on the plateau, they could see the Step Pyramid ten miles to the south at Saqqara. Then, in the clear desert air ten miles past the Step Pyramid, they could make out the shapes of the Bent and Red Pyramids. Sneferu had just been laid to rest in the Red Pyramid, but the Bent was a reminder to all of them that things can go wrong in pyramid building.

Hemienu had seen the problems that Sneferu, his father, faced when constructing his pyramid. Khufu was about forty years old when he became king; he didn't have infinite time for his pyramid to be completed and this figured in the design of the Great Pyramid.

Hemienu decided that there would be three burial chambers, each at different heights, in the Great Pyramid. The first, an underground burial chamber, could be completed in the first five years of construction, and would be used if the pharaoh died in his forties. The second burial chamber, inside the Pyramid, would be completed when Khufu was in his fifties and would be used if he died then. The third burial chamber, high up in the Pyramid, would be finished when Khufu was in his early sixties and would be the pharaoh's final resting place if he lived that long.

The location of these three chambers had been planned from the very beginning of the project. Their internal dimensions determined the volume of the Pyramid, how high it would be, the width at the base, and so on. When tourists look at the Great Pyramid, they often think that building a pyramid is merely stacking blocks on top of other blocks. Building the Great Pyramid was much more complex than that. There were two very different aspects to the Pyramid. There was the internal part with its rooms and passages, and there was the surrounding external, solid part. Each aspect required different materials and construction techniques.

As the team of builders looked out across the plateau, each man must have talked about his specialty—about the quality and quantity of stone, about where to cut a channel to the Nile, where to build the harbor, where to house the workmen, and so on. The most specific question they had to answer was exactly where on the plateau to build the Pyramid. It was clear that it should be on the east side of the plateau, the side nearest the river. That would allow materials and supplies ferried in on transport barges and off-loaded close to the building site, again reducing transportation time. This concern narrowed the Pyramid's final position to the eastern part of the Giza Plateau, but exactly where on the eastern part? A high elevation on the plateau was desirable so the Pyramid could be seen for miles, but Hemienu did not select the highest point on the plateau. Why not there, in the place of pride? Because Hemienu had a master plan for the plateau.

The highest point of the plateau is on the south side, where the gentle slope leads down to the southern outcropping of limestone that was going to be quarried. The slope forms a natural ramp, making it easy to transport blocks of stone up onto the plateau. But if Khufu's Pyramid were built at the highest point, right at the top of the slope, that would block the slope for future Giza projects using the southern quarry.

The Giza Plateau was large enough for six or seven pyramids, and Hemienu intended to create a 4th Dynasty burial ground for the royal family. By placing the Pyramid on the northeast corner, he left the slope on the south that led from the quarry open so two other large pyramids could be built in the future.

Hemienu thought on a large scale, both in space and time. The years ahead would provide many more examples of this genius's insights and skills, from organizing thousands of workers to selecting the perfect materials for each part of the Pyramid. Hemienu was a micromanager overseeing the largest building project in the history of the world.

Even before a single block had been quarried, Hemienu calculated that it would take twenty years to build the Pyramid. He knew it had taken ten years and seven months to build his father's Red Pyramid at Dashur, and that it rose at the rate of thirty feet per year. He knew this because he had watched it being built. We know it because of the work of the great German Egyptologist Rainer Stadelmann.

Stadelmann has spent his career studying Egypt's pyramids, and in the 1980s conducted a detailed survey of Sneferu's pyramids at Dashur. During this survey he noticed ancient writing on several blocks of the Red Pyramid on different levels. The inscriptions were from the time of the Pyramid's construction and gave the dates when the blocks were put in place. During the era of Sneferu, dates were often recorded in terms of the annual or biannual nationwide cattle census. Thus one block, on the thirteenth course of the Pyramid reads: "Day 14 in the second month of summer, in the year of the 15th cattle census." Another block, just three courses above reads: "30th day of the 3rd month of the season of inundation, in the year of the 16th cattle census."

Using the dates Stadelmann found inscribed on the blocks, another Egyptologist, Rolf Krauss, a specialist in chronology, concluded that it took nearly eleven years to build the Red Pyramid.[22] This estimate could be off by a year or so. Dates on blocks were usually inscribed at the quarry, so it is possible that blocks were stored there for some time before being transported to the Pyramid. Still, my bet is that Krauss's estimate is not far off. From the various dates inscribed on the blocks, he was also able to calculate that the Pyramid rose at a relatively constant rate of about thirty feet per year. At first this seems surprising because there are so many more blocks in the lower courses than the upper, so you might think the lower courses would take much longer to build than those higher up. However, it takes much more effort to raise a block all the way up to the top than it does to set a block in place on the bottom. It took eleven years to build the Red Pyramid, but Khufu's was going to be larger and its interior would be more complicated. Hemienu's calculations would have told him that twenty years was a reasonable estimate if nothing went wrong. One thing he was certain of, it would take thousands of workers to complete the project in only two decades, perhaps as many as 25,000 on the Giza Plateau.[23]

The population of Egypt was only a bit more than a million, so this workforce was a significant portion of the country's able-bodied men. At the beginning of the project the scene must have been a bit like America during World War II. Recruiters went throughout the country looking for healthy patriotic men. Families up and down the Nile were contributing husbands and sons for a national cause. Don't think of slaves; Hollywood got that one wrong. These were free men

The Giza Plateau was large enough to accommodate three large pyramids.

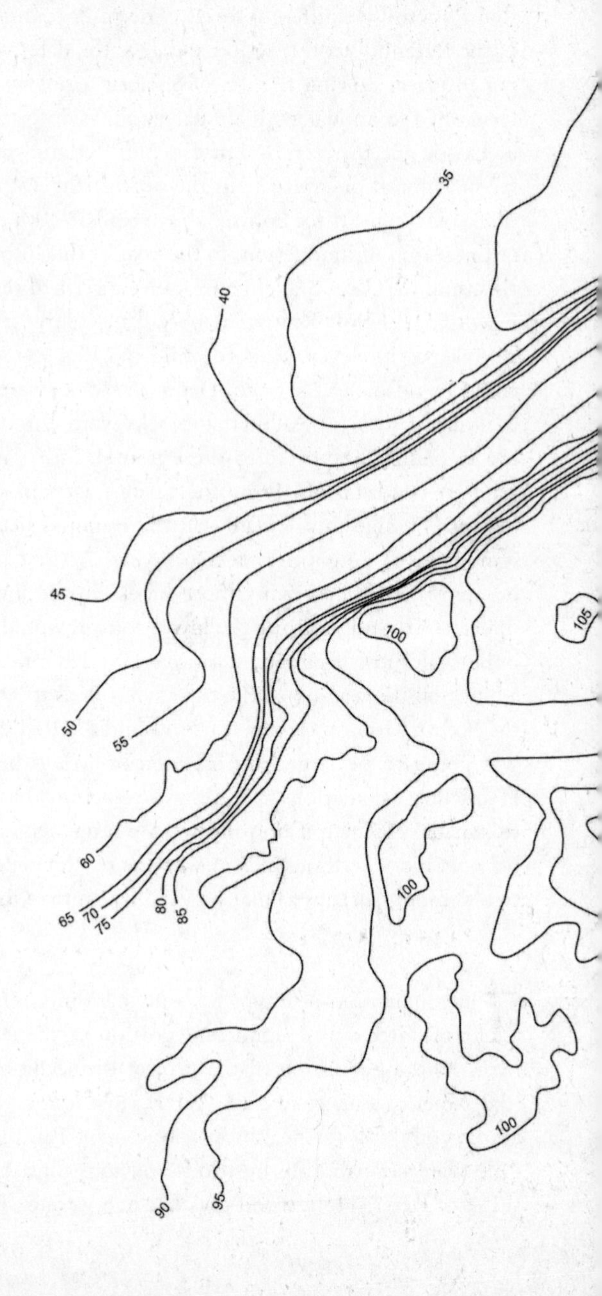

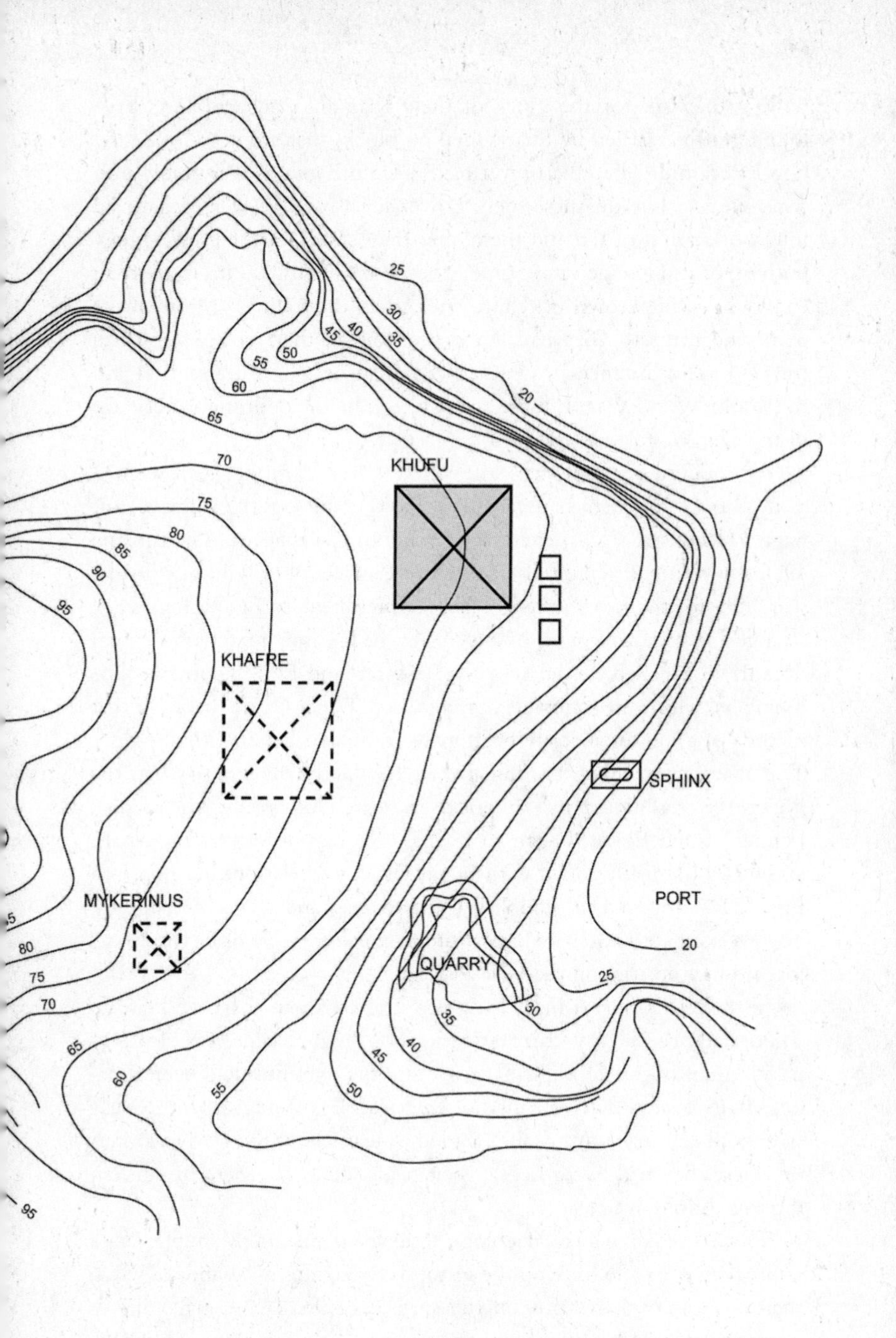
KHUFU
KHAFRE
MYKERINUS
SPHINX
PORT
QUARRY
25
30
35
40
45
50
55
60
65
70
75
80
85
90
95
20

willing to work for the glory of their pharaoh. Archaeologists have found graffiti written by quarrymen on blocks that are now inside the Great Pyramid. These hasty jottings in red ink give us a window into what life was like for these men. Lower-level workers were organized into work gangs of ten and the names they chose for themselves show both pride and joie de vivre. One group was "Khufu Is Pure"; another, "May the white crown of Khufu strengthen the sailing." Many of the men who came to Giza had never been out of their villages and had traveled great distances to a place they had only heard about. Again much like World War II, for many it would be the defining experience of their lives.

For modern Egyptologists, the workers were a mystery—it seemed as if all traces of them had vanished. How is this possible? They must have had houses, stores, everything a boom town needs. Then in the 1970s, a young Egyptologist from the University of Chicago began searching for the lost workers. Mark Lehner came to Egyptology via a strange and mystical route. As a teenager, he became infatuated with an organization called Association for Research and Enlightenment. The members, who were followers of the psychic Edgar Cayce, believed that records of a lost ancient civilization were hidden beneath the Sphinx. The association still exists, but had its heyday in the 1960s when the New Age was sweeping America. In search of arcane knowledge, Lehner visited Egypt, began to read Egyptology books and journals, and was soon disillusioned with Edgar Cayce's revelations but not with Egypt. He returned to school, learned to read and write hieroglyphs, how to excavate an archaeological site, and today is one of Egyptology's foremost authorities on the Giza Plateau.

In the 1970s, surprisingly little was known about the Giza Plateau. There had been plenty of investigations of the pyramids. The 4th Dynasty tombs of the nobility that surround the pyramids had been thoroughly excavated by German and American Egyptologists in the early twentieth century, but no one seemed to care about the lowly workers or where they had lived. Lehner began his search for them by surveying the entire Giza area.

The first stage of an archaeological survey is amazingly simple. Walk around the area and look on the ground. Even after thousands of years, there are often clues right on the surface of the site. Usually this is just pottery—the handle of a water jar, the bottom of a cup—but such fragments can provide information about who lived there. Just as our dinnerware styles have changed over the years, ancient Egyptian pots and jars had changed. A drinking cup used by workmen in the time of Khufu is different from one used a thousand years later by an artist painting Tutankhamen's tomb. Not only can the fragment tell you the period when a cup was used, it can tell you the status of the user. The cups of the elite were much finer than those used by workmen. Every excavation team has a pottery expert who can read these fragments of ancient lives.

When doing the surface survey there are other clues as well that can help narrow the area you are searching. Lehner knew that the workers' village wouldn't have been on the Giza Plateau itself. That was prime real estate, reserved for royal pyramids and tombs of the nobles. The workers would have lived just below the plateau, within walking distance, but in which direction? Ancient cities usually had enclosure walls for protection. Workers' villages also had walls around them, but for a different purpose. Usually there was only one entrance, and this enabled the village overseer to keep careful tabs on who came and went. When Lehner found the remains of a long stone wall just half a mile southeast of the Great Pyramid, he began excavating. Soon what became known as the Lost City began emerging from the desert sands.[24]

Everything a thriving village of hungry workers would need was there. The remaining walls of a bakery were only a foot high, but the site was littered with cone-shaped clay bread molds. These molds were once filled with dough, stacked high in an oven, and baked. Any Egyptologist can tell you that if you find the bakery, the brewery can't be far away. Egyptians drank beer and you need yeast to brew beer. It was a simple matter for the brewer to get his yeast from the baker. The bakery and brewery weren't Lehner's only finds. Across town he found a granary where huge amounts of grain were stored. It was a spacious courtyard filled with seven round mud brick silos, each about eight feet across. There are probably more nearby, beneath a modern soccer field where today boys play in the shadow of the Pyramid their ancestors built.

When Lehner found the remains of the butcher shop, another expert on the team sprang into action. Paleozoologists can look at an ancient animal bone and tell you whether it is the knee cap of a goat or the ear bone of a pig. If you are excavating an ancient city and want to know what the inhabitants ate, these people are crucial. The analysis of the bones found near the butcher shop revealed that the workers ate quite well. They were supplied with beef, sheep, goat, and occasionally pork. This was not the usual poor man's diet. The typical laborer rarely ate meat; his basics were bread, beer, and onions. If the pharaoh's workers were going to be doing strenuous labor, they would need plenty of protein and calories; they had to be fed well. Because it was difficult to keep meat fresh in ancient Egypt, much of these supplies were probably delivered on the hoof and slaughtered daily as needed.

There were also administrative buildings where scribes and officials kept track of all the goods supplied to the workforce. Lehner even uncovered a large house he calls the Manor, where a high official who oversaw the workers' village lived. Of all the buildings uncovered in the thirty years he has been excavating the Lost City, the strangest, discovered in 1999, is known as the Gallery. It's a 100-foot-long building whose main feature is a mud brick partition running right down the middle, dividing the gallery into two sections each about six feet wide. It looks like Siamese bowling alleys. As excavations continued, more and more galleries were discovered throughout the Lost City. For now, the best bet is that they functioned as barracks, sleeping quarters for workmen, who would stretch out side by side, looking very much like two rows of sardines in a can.

Lehner's three decades of excavating the Lost City has made the workmen come alive. We know what they ate, how they slept. We know there were other areas of the city where weaving took place, and where craftsmen hammered out copper chisels needed to quarry and shape the Pyramid's blocks. It was a hive of activity and everyone was working for the glory of the pharaoh.

When this army of workers started pouring into Giza, their first jobs were to build their houses, bakeries, breweries—everything a city of 25,000 needed to sustain itself. All the buildings were simple affairs made of mud bricks. No expensive stone could be wasted on mere workmen. There was little furniture in these modest houses, just a chest woven from palm fronds to hold clothing and some reed mats to sit and sleep on. Some of the workers were put to work digging the four-mile canal to the Nile so supplies could be floated right up to the work site. Although five-sixths of the Pyramid's blocks would come from the Giza quarries, more than a quarter of a million tons of special stone still came from off-site quarries and had to be transported to Giza. Hauling sixty-ton granite blocks on sleds from the Nile over four miles of uneven sandy terrain is almost unthinkable. It was much easier to float them down the Nile and through the canal to the site. When the Egyptians did transport blocks of stone over land, they didn't use wheels. One reason is that wheels sink into sand, but another is that they didn't have a metal strong enough for axles to support blocks weighing several tons. So when the blocks came off the transport barges at the site's harbor, they were already on their sleds and were hauled a short distance to the Pyramid. Some of the sixty-ton stones were twenty-four feet long and four feet thick, and were placed on huge thirty-foot-long sleds made of thick cedar beams. Then with an overseer standing on top of the granite block calling out instructions, hundreds of workers pulled at the ropes to haul the block.

There were two off-site quarries; the closer one was eight miles upstream from Giza across the river on the eastern bank. The limestone at Giza was yellow, rough, uneven, and pocked with fossils, suitable only for the Pyramid's inner fill that no one would ever see, not fine enough to cover what would become the Pyramid's smooth, white exterior. For this purpose, nearly a quarter of a million tons of pure white covering would be cut from a 500-foot glistening mountain of limestone. The stoneworkers used four-inch-wide copper chisels and wood mallets. Copper is soft and dulls quickly, but fortunately limestone is also soft and splits along straight lines, so copper was adequate for the task. However, there must have been a small army of unskilled workers camped at the site whose job was just to sharpen the chisels as they dulled.

Limestone has an unusual property that makes it ideal for quarrying. It is soft when first quarried but hardens when air hits it. Thus it was easiest to finish the fine blocks on the site when they were still soft. The average facing block weighed more than two tons and was three feet high, so several men could polish it at one time. Today the site where these blocks were quarried is called Tura. During World War II, the British stored ammunition in the tunnels excavated by Khufu's quarry workers. After the war, the quarry was reopened for modern building stone and the ancient tunnels and traces of the workers were destroyed.

The second off-site quarry Hemienu needed was 500 miles south of Giza, on Egypt's southern border at Aswan. Here a very different kind of stone was found—granite. One hundred times harder than limestone and extremely difficult to extract and shape, it created a great challenge. Limestone couldn't be used to span large spaces; it isn't strong enough, but granite is. Up until the time of Hemienu, granite was used sparingly because it is so difficult to work. But Hemienu decided that his king's burial chamber inside the Pyramid would be constructed entirely of Aswan granite. It was needed purely for structural reasons. Khufu's burial chamber was going to be high inside the Pyramid, with hundreds of thousands of tons of stone above its roof. Somehow, the ceiling had to support this enormous weight above it. In the earlier pyramids at Dashur, a corbelled ceiling solved this problem, but Hemienu had another plan: he would do it without corbelling. We don't know who made the decision to abandon corbelling. Perhaps the king himself decided he wanted something even greater than his father's burial chamber. Maybe it was Hemienu's idea. We can't be sure, but we do know that the goal was a burial chamber with a flat ceiling. This would create a more open space than a corbelled ceiling that got smaller and smaller toward the top. Such a room would require granite, lots of it. Hemienu's calculations would have indicated that more than 3,500 tons had to be quarried, shaped, and shipped down the Nile to Giza. Since he knew both exactly how high up in the Pyramid the burial chamber would be and also the rate at which the Pyramid would rise each year, he knew the granite for the burial chamber would not be needed until the twelfth year of construction. He also knew it would take years to free that much granite from the Aswan quarry, and that work had to begin immediately.

Granite was too hard to be worked using copper, the only metal available for chisels. Every block would have to be pounded out and shaped with hard dolerite rocks, some weighing sixteen pounds. Today, 4,500 year later, the ancient quarry is still littered with hundreds of round, black rocks. Hemienu's workmen repeatedly lifted and dropped these pounders on the granite, chipping away only millimeters at a time. Using this laborious process, they ultimately fashioned forty-three granite beams twenty-four feet long, weighing from thirty to sixty tons each. Add to this hundreds of smaller blocks, many four-foot cubes, but still weighing many tons each, and it seems an almost impossible task with only stone tools. But Hemienu had done his calculations. If the men were dispatched to Aswan now, at the beginning of

The ancient dolerite pounders used to quarry granite.

the project, the granite blocks should be ready for the pharaoh's burial chamber when they were needed, just as the Pyramid approached 150 feet in height.

Probably five hundred men were dispatched to open the new quarries at Aswan. In twelve years they would provide more granite than Egypt had used in its entire history. An even larger number of men were sent to Tura to begin quarrying the fine limestone casing blocks that would make pharaoh's Pyramid glisten in the sunlight. From the earliest stages of the Great Pyramid's construction, Hemienu had four quarries simultaneously in operation: 1) The Giza Plateau itself was quarried around the Pyramid's thirteen-and-a-half-acre base to provide limestone for the inner core. 2) Just south of the plateau, the southern quarry supplied more low-grade limestone for the inner core. 3) The Tura quarry, just a few miles to the south and across the river, provided the fine, smooth limestone for the Pyramid's outer casing. 4) The Aswan quarry, 500 miles south, yielded all the granite for the interior chambers of the Pyramid.

In a period of twenty years, these four quarries would supply more than two million blocks of stone to construct the Great Pyramid, on average, more than 100,000 blocks a year. The men probably worked a ten-hour day, which would mean that a completed block was quarried, transported, and pushed in place every three minutes—365 days a year for twenty years! One wonders which is more remarkable, the Great Pyramid's construction or the social organization needed to bring about that construction.

Boats had to be built to transport hundreds of thousands of massive stone blocks. Three thousand five hundred tons of granite was shipped from the Aswan quarry during the first twelve years of construction. That's about 300 tons a year that had to make the thirty-day journey north to Giza. Far more than that had to be shipped from the Tura quarry, but that was only an eight-mile haul. The transport barges had to be sturdy enough to support the heavy blocks and this required stout timbers, but Sneferu, Khufu's father, had already solved that problem. When Sneferu was building his pyramids, he sent an expedition to Lebanon to trade papyrus, gold, and finely crafted objects for the massive cedars of Lebanon. Khufu must have done the same for his pyramid fleet.

Although the Pyramid is huge, and the quantities of stone, men, and supplies seem overwhelming, sometimes things are simpler than they appear. When I started thinking about how many boats were needed to transport the fine white limestone for the facing of the Pyramid from the Tura quarry, I had visions of the Nile clogged with barges. After all, hundreds of thousands of tons of stone were needed to face the four sides of the Pyramid. So let me ask you a question. How many boats had to be built to transport all that stone? One of my students answered the question very cleverly.

Once a year I teach an Egyptology elective course at the Webb Institute for Naval Architecture. An amazing place, one of the finest schools in the world for naval architecture, named after William Webb, who during the Civil War designed the *Monitor*, the North's famous ironclad battleship. Webb went on to become America's top ship designer; when he died he left his fortune to found the Webb Institute. He wanted the best for the best, so every student is on a full scholarship, and since only twenty are accepted each year, it is highly competitive. You may have seen the school. It is an old Edwardian-style mansion on Long Island Sound—and it was Wayne Manor in one of the *Batman* movies.

Anyway, since the kids are so bright, I decided to ask my class the question I've asked you: How many boats were needed to transport the facing stones from the quarry to the Pyramid site? I wanted the class to think about the problem and give me their answers at our next meeting, but no more than thirty seconds had passed when one student raised his hand and said, "Two." Incredibly, he's probably right. He knew that the facing blocks were about three feet deep and had calculated how many tons of stone were needed for a pyramid 480 feet high with a 756-foot base. Then he calculated the number of work days in twenty years. Let's say it's 300 days per year, so during twenty years there would be 6,000 work days. A boat could easily make the eight-mile voyage from Tura to Giza in a day, so if a boat took forty tons and you had two boats you could easily transport 250,000 tons of stone to the site in twenty years. That's predicated on the assumption that each boat hauled forty tons. We must also remember that the Great Pyramid was not the only monument under construction on the Giza Plateau; there was an entire funerary complex. There were three small pyramids for Khufu's queens, a mortuary temple, and a causeway, all requiring limestone. This would have raised the total amount of limestone used and raised the number of transport boats needed.

We know that the transportation of all this stone was well within the capabilities of Egyptian boats because of a scene carved on Queen Hatshepsut's temple about 1,000 years after the Great Pyramid was constructed. We are shown the moving of Hatshepsut's two great obelisks from Aswan to the Karnak temple. Each obelisk weighed about 250 tons, and both were placed on a single barge.[25] Thus a forty-ton barge for moving the facing stones would have posed no difficulties for Khufu's workers. Even if they decided to have smaller boats capable of moving only five tons or two blocks, we are still talking about only a couple of dozen boats. Granted, you'd need to transport more stones in the earlier years than in the later years because the pyramid is larger on the bottom that the top, and you would also need extra boats to replace damaged ones, but still it is a manageable task.[26]

Harbors, ships, houses, and canals were only some of the resources needed to build Khufu's Pyramid. Miles and miles of thick rope had to be woven to haul the millions of blocks on sleds to their final positions in the Pyramid. In the limestone quarries, thousands of copper chisels were needed to free the blocks from their matrix and cut them into the desired size and shape. Over the course of the next twenty years, more than 250 tons of copper would be needed for the pharaoh's chisels.[27]

Expeditions of miners were dispatched to the Sinai Desert, Egypt's only source of copper. The journey to the Sinai was long and difficult. This was foreign territory inhabited by "barbarians." The miners and their equipment, protected by the pharaoh's soldiers, trekked through Egypt's eastern desert to the banks of the Red Sea and then boarded transport vessels to cross to the Sinai. Then the party of fifty or so miners and their guards mounted donkeys for the nine-day journey to the mines.[28] The ore they chipped out of the mountain was crushed and then heated in kilns so the copper would run out. In pure form, the copper ingots were taken back by donkey to the transport ships and across the Red Sea. Once on the other side, it was back to the donkeys to pack it across the desert to the Nile. Finally the copper was shipped to the Pyramid site to be cast into the tools urgently needed by the quarrymen.

Mortar was another important ingredient in the Pyramid. The Pyramid is much like a dam—the exterior surfaces are smooth and polished, but the interior is mostly rubble. When the Pyramid was built, first the smooth white Tura facing blocks were slid in place, then behind them were a few rows of backing blocks. Then the inside of the Pyramid was filled with irregular rough blocks. As the crude interior blocks were set in place on the Pyramid, the spaces between them were filled with mortar. From a quarry near the Giza Plateau, half a million tons of gypsum were mined and transformed into mortar for Khufu's Pyramid.[29] This involved digging up the gypsum (calcium sulfate), burning it so that it loses about three-fourths its water, and then pulverizing it. When water is now added, it recombines to make something very similar to our modern plaster, setting and becoming very hard.[30] The list of jobs tackled by Hemienu's pyramid builders seems endless, and much of the success of the project was intimately linked to a huge supply chain, the first ever of such magnitude.

As the workers were pouring into the Giza Plateau, building their houses, digging canals, heading off to the Sinai for copper, and building barges to transport stone, Hemienu would have been undertaking the task of orienting the Pyramid. After it was decided that the Pyramid would occupy the northeast corner of the plateau and have a square base of thirteen and a half acres, it was determined for religious reasons that the Pyramid should be oriented to the four points of the compass. The stars seem to rotate around a fixed point in the sky. Today that point is the North Star. In the time of the Great Pyramid, it was different, but as the other stars rotated, their northern point was also constant. The deceased pharaoh was associated with this unchanging, eternal northern point; thus the entrance to the Pyramid would face due north toward that point. Orienting the sides of the Pyramid to the compass points would not have been a difficult task. The ancient Egyptians were skilled surveyors. Each year when the Nile overflowed its banks, boundary markers were washed away; when the water receded and the land emerged, everything had to be resurveyed. With centuries of surveying practice, the ancient Egyptians became quite good at it.

There were several possible methods by which they could have determined true north. One method was solar. At noon, a perfectly vertical staff called a gnomon will not cast a shadow because the sun is directly overhead. Mark the shadow it casts five minutes before noon, then mark it five minutes after noon, bisect the angle and the line of bisection will point due north. A series of gnomons would provide greater accuracy when all their lines were connected.

A second method used the stars. A wall about five feet high is built on the site, creating an artificial, perfectly level horizon. In the evening, facing in the general direction of north, an astronomer-priest would observe the passage of the stars across the sky. He would mark the wall where a given star first rose above it. Later that night, he would mark where the star disappeared behind the wall. The midpoint between these two points is north. This observation could be repeated many nights with different stars to obtain an exact reading for north.[31]

Another astronomical method of determining north has recently been suggested. This involves the careful observation of two stars, Kochab and Mizar. When one was directly above the other, the line connecting them indicated true north.[32] Whichever technique the ancient Egyptians used, it worked. The sides of the Pyramid are remarkably aligned to the four points of the compass.

Hemienu understood that with something as large as the Great Pyramid, precision is crucial. Being off by an inch at the base could cause a deviation of yards at the top. Thus the base had to be as close to perfectly level as possible—and it is. At the base of the Pyramid, each side stretches 230 meters but never deviates from level by as much as an

To orient the Great Pyramid's four sides to the four points of the compass, astronomer-priests may have observed the rising and setting of stars behind an artificial horizon.

inch. We are not certain exactly how Hemienu achieved such accuracy, but it has often been suggested that the principle was the same as using a modern carpenter's spirit level, the kind with a bubble of liquid in it. A thin trench was cut around the perimeter of the Pyramid and filled with water. The level of the water was then traced on the wall of the trench and that was the starting base line. The trench could be filled in up to the base line and the rest of the perimeter leveled in accordance with the line. Once the perimeter was level, the first blocks were almost ready to be put in place.

The base of the Pyramid covers approximately thirteen and a half acres, about ten football fields, but Hemienu didn't level the entire area. Rather, he left an outcropping of limestone twenty-one feet high still attached to the bedrock in the middle of the base. Thus the first twenty-one feet of the middle of the Pyramid did not have to be built out of blocks; blocks were only needed for the perimeter. Tourists who visit the Pyramid don't realize this because the stonemasons sculpted the bedrock to look like blocks. The average Pyramid block is a three-foot cube. Some of the exposed "blocks" appear to be thirty feet long. They are not really blocks, just carved bedrock, so other blocks could be stacked on top of it. The outcropping comprises approximately 10 percent of the Pyramid's mass—a substantial saving of labor and materials. With the perimeter leveled and the first twenty-one feet of the center of the Pyramid already rising from the quarry, the actual building was ready to begin. This is where planning was crucial.

There is a tendency to think of the Great Pyramid as simply a huge building block construction. Start hauling blocks up a ramp, push them in place, and repeat the process until you come to a space where you want a chamber. Build around that space and then continue placing the giant blocks until you reach the top of the Pyramid. Such building techniques will not get you the Great Pyramid of Giza. The construction and engineering problems to be solved are far too complex for such a simpleminded approach. For example, the techniques for installing blocks at the bottom of the Pyramid cannot be used at the top. At the base you merely push the stones into place, but when you near the top of the Pyramid it gets tricky—you have to raise the stones 450 feet. Also, each chamber inside the Pyramid is designed differently and requires different construction methods.

Because there were so many different aspects to the Great Pyra-

What appear to be very long blocks of stone are really part of the Giza Plateau bedrock that has been shaped.

mid, it was an extremely difficult project to plan, but by the time Khufu became pharaoh, Egypt had been building pyramids for a hundred years. On the Egyptian scale of things, a century doesn't seem like much, but that is a long time to gain experience and to perfect construction techniques. When Hemienu began planning his pharaoh's Pyramid, there was quite a history of pyramid construction to draw upon, though he intended to go far beyond anything that had ever been done or imagined.

The Underground Burial Chamber

Giza, Egypt, 2584 B.C. (Year 5 in the reign of Khufu)

During the first few years of building, all problems in the work plan were ironed out. Where the men would live, how they would be fed, and who would make up the various work gangs or teams were all decided. Tremendous social organization was needed, but such issues had been figured out during the construction of the earlier pyramids.

While the bedrock was still exposed, masons began carving out the descending passageway that led to the underground burial chamber. This chamber was to be used only if Khufu died during the first ten years of construction. Because the descending passage was never intended for heavy traffic—just the pharaoh's sarcophagus—it is rather small, just large enough for one person to walk around or move in hunched over. It must have been carved by a single master stonemason, as there is no room for two people to work in it. Of all the chambers and passages in the Great Pyramid, it is the most precise. An engineering marvel, the exactly cut rectangular tunnel descends 230 feet, pointing due north and never deviating by more than a quarter of an inch. At the early stages of the Pyramid's construction when the subterranean chamber was being carved out, this passage served as the highway that daily brought the workers to their jobs underground.

Once the descending passage was completed, a team of workers began carving out the burial chamber. At first only one man, working at the end of the passage, could carve the beginnings of the subterranean chamber into the bedrock. Then, when there was room for a coworker, a second mason joined him and they chipped away at the limestone until the small chamber could accommodate a third worker, and so it went until a team of more than a dozen were working there. Just as in all excavated tombs, they began at the ceiling and excavated downward, assisted by gravity. If you carved a tomb from the bottom up, you would always have to swing your mallet upward and fight gravity.

The men who excavated the underground chamber didn't need great skills. They were really just roughing out the twenty-six-by-forty-three-foot room. As they worked and the chips accumulated, even less skilled workers carried the chips in baskets up the descending passage to the surface. Later, more skilled craftsmen could put the finishing touches on the king's burial chamber. But the skilled craftsmen were never needed. As the Pyramid progressed the pharaoh was clearly in good health. Hemienu could now move on toward a burial chamber high up in the Pyramid.

After the first few years of work, the Pyramid had risen only about twenty of its 482 feet, but the outcropping of bedrock left in place had now disappeared inside the Pyramid. As the Pyramid grew, tier by tier, it was not perfectly level. Only the outer-facing blocks, already carefully dressed, and those immediately behind them were set in place with precision. Inside this thin perimeter, rough blocks of various sizes were crudely pushed into place and the spaces between them filled with rubble and mortar. Every thirty feet or so, the Pyramid was leveled carefully and then the construction continued. As Hemienu prepared to build the second burial chamber, raising blocks efficiently became essential. In the very first stages, it was a simple matter to just push blocks into place, but when the Pyramid rose past 150 feet or so, raising the two-and-a-half-ton blocks became a real chore.

The stone columns at the entrance to the Step Pyramid were carved to resemble bundles of papyri tied together.

The Bent Pyramid was Egypt's first great architectural disaster.

The pyramid of Meidum looks more like a fortress than a pyramid because in the Middle Ages its stones were quarried for other buildings.

An 1880s photo of the Great Pyramid.

FOLLOWING PAGES:
Computer reconstruction of the Great Pyramid after 5, 14, 15, 19, 20, and 21 years of construction.

5 YEARS

The Pyramid is just a few courses high and the white bedrock core inside the Pyramid is still visible.

14 YEARS

The Burial Chamber is now being constructed and the external ramp used to build the lower courses of the Pyramid has reached its maximum height.

15 YEARS

The Burial Chamber is under construction and a kind of step pyramid is built at 60 feet. The external ramp is still in place.

19 YEARS

The Burial Chamber and relieving chambers have been completed, and the external ramp used for the lower courses has been removed and the materials have been reused to complete the upper part of the Pyramid.

20 YEARS

The Pyramid nears completion.

21 YEARS

The Pyramid is complete with all its gleaming white facing stones in place.

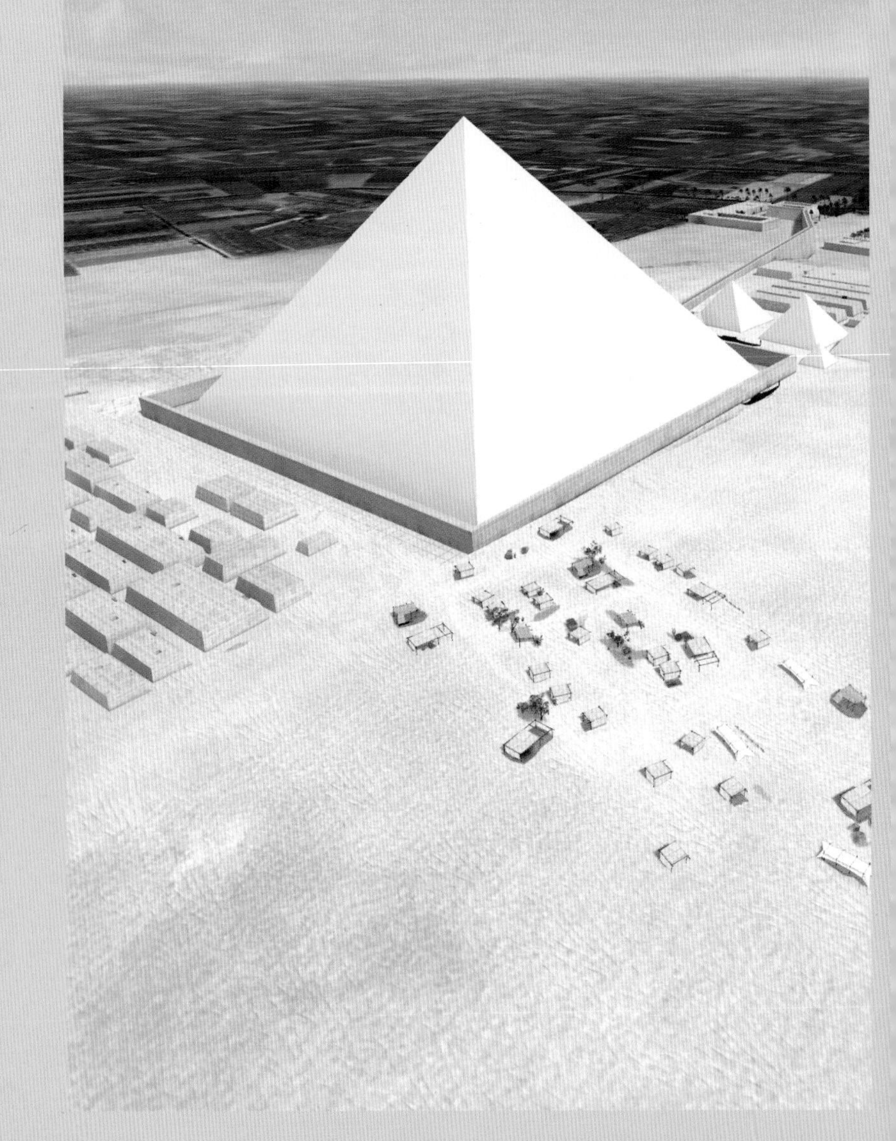

For the early stages, a simple exterior ramp would do just fine. Up to about 200 feet a ramp is neither too massive nor too long to be impractical and undoubtedly Hemienu built the first 200 feet or so with a single ramp. The terrain dictated that the ramp be on the south side of the Pyramid; it was the only side of the Giza Plateau without a steep drop-off. The south was also where the quarry was, so this reduced the distance the blocks had to be hauled. At the beginning stages of the Pyramid's construction, the ramp had to be quite wide, more than 100 yards, to accommodate teams of workmen simultaneously pulling blocks up. At that point, one block was being pushed into place every three minutes. From the start, the ramp was built at its maximum length, stretching a quarter of a mile south toward the quarry. To make sure that a constant stream of blocks was always moving, the ramp was probably divided into two parts. One part was used for hauling and for the workmen returning back down. Meanwhile, the other half was being raised in readiness for the next course of stones. Once an additional level of stones was in place, the workers started hauling up the other side and the side they had just abandoned was raised to be ready for the next level. Thus half the ramp was always being built up while the other half was in use.

The majority of the blocks that went up the ramp were rough two-and-a-half-ton filler blocks extracted from the southern quarry. Some were much larger. When the Pyramid was sixty feet high, twelve huge twenty-ton limestone blocks were hauled up the ramp. It was time for construction to begin on the second burial chamber, usually called the Queen's Chamber.

Compared with the unfinished underground burial chamber, the eighteen-foot-square second chamber is quite small. It appears almost as if Hemienu knew the king would live long enough to be interred in the third burial chamber that would be built higher up inside the Pyramid. In the Queen's Chamber, a roofing technique was tested that would later be crucial in the third burial chamber.

The ceiling is the weak point of any chamber inside a pyramid. It must support the enormous weight of the pyramid above it, and no beams—not even granite—can bear that kind of load. In the earlier pyramids at Meidum and Dashur, there were no beams; corbelling was the solution. But the third burial chamber was going to have a flat ceiling, the first ever inside a pyramid. Somehow, the weight had to be taken off the ceiling beams or they would crack.

Rather than corbelling the walls inward as they rose to the ceiling, the walls of the second burial chamber go straight up like traditional walls, but then they are capped by six pairs of huge white limestone rafters that meet at a 120-degree angle. These rafters distribute the weight of the Pyramid above them into the solid body of the Pyramid below. The rafters were put in place sometime after the tenth year of construction, and an ascending passage leading to the second burial chamber was also built. So now two passageways led from the Pyramid's entrance to two burial chambers—the unfinished one beneath the Pyramid and the second burial chamber inside the Pyramid. Like the first burial chamber, the second is incomplete, again a sign that Khufu was healthy and expected to live long enough to be buried in the third chamber, even higher up in the Pyramid.

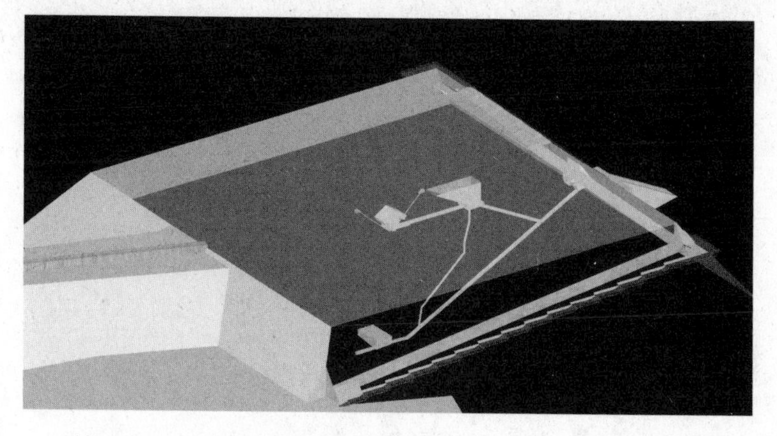

After ten years of construction, two burial chambers and two passageways had been completed inside the Great Pyramid.

The Queen's Chamber was the first to use rafters to support the weight of the Pyramid above.

Modern Tomb Raiders: The Search for Hidden Chambers

The Great Pyramid, 1986

The vast majority of tourists who visit the Great Pyramid never go inside; many don't even know it's possible. Group tours to Egypt are a whirlwind affair—"Not much time, we have to keep moving." Usually a morning visit to the Giza Plateau involves a walk around the Great Pyramid as the guide explains how tall it is, who built it, and how many stones were used. That is followed by a drive to a spot where you can take a "panoramic picture" of the Giza pyramids, and then a drive to see the Sphinx and the Valley Temple of Kephren. If there are a few minutes to spare, and the traveler is adventurous, there are plenty of camel drivers eager to take him for a ride. The special ticket to go inside the Pyramid costs 100 Egyptian pounds (about twenty dollars) and only 100 are sold each day, to keep the humidity constant inside the Pyramid. Thus, it is the rare tourist who sees the inside of the Pyramid and almost everyone who does is confused by what he sees.

To the first-time visitor, the Pyramid seems to be a maze of corridors and rooms with no clear purpose. The best way to make sense of the Pyramid is to remember that there are three burial chambers and a passage leading to each. 1) There is a descending passage that goes into the bedrock to the subterranean burial chamber. 2) The chamber erroneously called the Queen's Chamber higher up in the Pyramid is reached via the ascending passageway. 3) The last burial chamber, the King's Chamber, highest up in the Pyramid, is reached by a passage called the Grand Gallery. Three burial chambers, three passages; sounds easy but once you're there it's overwhelming.

For centuries there have been rumors of hidden chambers filled with treasures Khufu intended to take with him to the next world. There have been plenty of failed scientific attempts to find them, so by now you would think that everything is known about the Pyramid. Not quite. Incredible as it seems, in 1986 two French tourists visiting the Pyramid discovered evidence for a chamber that had remained hidden for 4,500 years.

The two men, Gilles Dormion, a design technician, and Jean-Patrice Goidin, an architect, were hunched over in the low horizontal passage that leads to the Queen's Chamber when they noticed something

When finally completed, the Great Pyramid had three burial chambers and three major passageways.

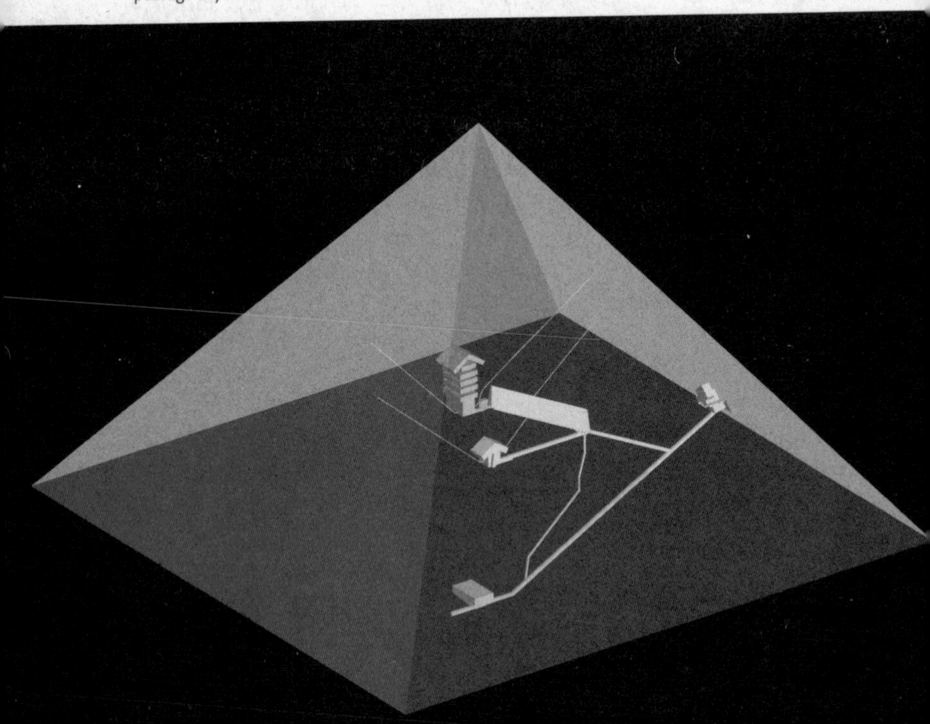

strange. The passage is about four feet high and 100 feet long—not for the claustrophobic—and is constructed of polished limestone blocks. In any culture, when you build with blocks or bricks, you normally don't stack one block directly on top of the other. That would create unstable tall columns of blocks. Just look at any brick wall and you'll see the familiar pattern of a brick on top resting on the two bricks beneath it. This is also the way the Great Pyramid was built. Go through the passages and chambers and you will see one stone resting on two beneath it. Our Frenchmen were experienced at building and they knew this, too. As they duckwalked through the low passage, something caught their eye. A section of one of the walls was constructed the wrong way: the blocks were stacked up directly one on top of the other so that the joins formed plus signs. This is the only place in the Great Pyramid where blocks had been laid this way. Why? As visions of hidden treasure danced in their heads, they wondered, could there be a hidden room behind the special wall?[33]

An unusual pattern of stacked blocks led to the 1986 discovery of a previously unknown chamber. The metal screw heads are stoppers for the drill holes.

The two men returned to France with dreams of excavating and finding Khufu's treasures. It might seem rather fanciful for two tourists to even be thinking about excavating in the Great Pyramid, but this was the 1980s and regulations about excavating were much more lax than they are now. Today you must be affiliated with a university, have academic credentials, submit a detailed proposal of exactly what you want to do and how you intend to do it, and have an equally well-documented team behind you. But in the 1980s, if you were well connected and didn't look like you were going to damage anything, you had a chance. Dormion and Goidin were well-respected professionals who knew other well-respected professionals, and soon they had put together a team called Opération Khéops.

They were amateurs, but they planned their project well. They intended to study the Pyramid nondestructively—without moving any blocks. To do this they needed some high-tech help. They enlisted the aid of a company that used a technique called microgravimetry. The principle is simple, the application is not so simple.

Anything with mass has a gravitational attraction. As is well known, the moon's gravitational attraction to Earth causes the tides. The gravitational force of the stone blocks in the Great Pyramid can be detected with a very sensitive device called a microgravimeter. Imagine one small section of the Pyramid that is about the size of a house. We can use the microgravimeter to measure the gravitational force of this area. We can compare this force with that generated by a second section of the Pyramid of exactly equal size. What if the second section has a smaller gravitational force? What would that mean? It's the same "size" as the first section, but it must have less mass. Why? One reason could be that inside the area is a hollow space, an unseen chamber. There are fewer stone blocks in this section, less mass, and thus a weaker gravitational force. This is the principle of microgravimetry—detect minute gravitational anomalies in the Pyramid and you might find a hidden chamber.

With the help of some high-placed friends, the two Frenchmen obtained the backing of the French Ministry of Foreign Relations and submitted their formal proposal to the Egyptian Antiquities Organization. There are two very different kinds of permissions one can apply for. One is to excavate, in which you're requesting permission to dig in the ground, shift sand, or move blocks of stone. Here, if you don't know what you're doing, you can damage the tomb or temple you are excavating. The second kind of permission is to survey a monument. In this kind of project you merely want to map, diagram, and record the monument. Opération Khéops's application was for a survey. The Frenchmen proposed the most comprehensive search for hidden chambers ever. The proposal was accepted and soon Opération Khéops was off to Egypt with a load of high-tech equipment.

Opération Khéops took thousands of readings with the microgravimeter, not just where they suspected the hidden chamber might be, but up and down the entire Pyramid from all four sides. While they were working, their architects drew maps of every wall in every known chamber and passage in the Great Pyramid, producing the most detailed plans of the Pyramid ever created.

When Opération Khéops returned to France, a Cray 1 supercomputer analyzed the thousands of measurements taken by the microgravimeter and printed out the equivalent of hundreds of X-rays of the Pyramid. Sure enough, the computer analysis showed some kind of cavity behind the west wall of the passage leading to the Queen's Chamber, just where the Frenchmen had noticed the unusual stacking of the blocks.[34] The new finding was so intriguing that the Egyptian Antiquities Organization permitted the French to drill three small holes through the wall to see if they could reach the hidden chamber. After drilling the first hole for nearly nine feet, they found nothing, just solid blocks of stone. They changed the angle of the second hole, thinking that perhaps they had drilled above or to the right or left of the hidden chamber, just missing it. Again, no luck. The third hole was their last chance; they would be permitted no more drilling in the Pyramid. Once again changing the angle, and after nearly nine feet of solid rock, they broke through into a room filled with fine sand. What had Hemienu done?

At first their instinct was to drill deeper, past the sand to see if there was something beyond. But drilling deeper in such a narrow passageway would be difficult and it was not clear if the Egyptian Antiquities Organization would permit more holes in their Pyramid. So the team returned to France to retool their instruments for a more detailed microgravimetric survey of the newly revealed chamber. As they were planning their return campaign, a Japanese team came up with an even better way of seeing what was behind the wall—ground penetrating radar (GPR). They too obtained permission from the Egyptian Antiquities Organization to survey the Great Pyramid. Unlike the French team, the Japanese were affiliated with a university, Waseda in Tokyo, that had a tradition of excavating in Egypt. Their leader, Dr. Sakuji Yoshimura, was a professional Egyptologist, so the Japanese had several advantages over the amateurs.

From January 22 to February 9, 1987, the Japanese conducted their survey using GPR. With ordinary radar, electromagnetic waves are sent out by a transmitter, the waves are reflected by the target, and a receiver interprets the reflected waves. With ground penetrating radar, high-frequency radio waves are transmitted into the ground and the receiver interprets the returning signals to determine if there are any anomalies in the ground. GPR is used extensively in building projects to make sure the foundation for a building isn't above a natural cavern. The Waseda team sent radio waves through the west wall of the corridor leading to the Queen's Chamber, and confirmed a passageway parallel to the west wall that is filled with sand. The Japanese were not allowed to drill holes, so the contents of this space remain unknown.

Taking the French team's discovery of the sand one step further, the Japanese analyzed it under a microscope and learned that it is not local. Sand at Giza and Saqqara is mostly calcite and other minerals. This sand is almost pure quartz and the grains are much larger than those of local sand.[35,36] What was so special about this sand that caused the ancient builders to import it? And what was it doing inside the Pyramid? No one knows the answer, but there are two points to be learned from the French and Japanese expeditions to the Great Pyramid. The first is that there are still discoveries to be made inside the Pyramid. The second, as we shall see later, is that without realizing it, the French may have discovered the internal ramp.

Athens, Greece, September 1988

Soon after the French team's discovery, members of the team presented their findings at the annual international conference of the IAEG (International Association for Engineering Geology and the Environment). A high-tech conference, it highlights how new technologies can assist geological surveys, prospecting, and other ventures. In 1988 the conference was held in Athens, and H. D. Bui, the team member responsible for analyzing the data, presented the paper. Bui is what scientists call a heavy-hitter. He is a distinguished member of both the French and European Academies of Science. A slightly built man of fifty who speaks with a gentle voice and a tone of authority, he has a reputation for rarely being wrong. As he began his lecture, he stressed the potential usefulness of microgravimetry to archaeology, and then went on to describe the discovery of the chamber behind the west wall and the sand that it contained. He explained that aside from this finding, their project had nothing to add to the construction of the Great

Pyramid. He casually mentioned that his analysis of the data produced one anomaly that didn't make any sense, but it should be mentioned for the sake of completeness. The anomaly showed "zones [that] are not correlated horizontally and can suggest various building arrangements . . . spiral constructions . . ."—ramp-like areas. No one seemed interested in this observation and it passed without comment. Most of the scientists were geologists, not particularly interested in archaeology.

The Grand Gallery

Giza, Egypt, 2579 B.C. (Year 10 in the reign of Khufu)

Around the tenth year of construction, the rafters of the Queen's Chamber were slid in place and Hemienu's focus shifted toward building the third and last burial chamber. However, before constructing the King's Chamber, Hemienu built another room, a room so strange that no one has ever explained its purpose.

The Grand Gallery has puzzled Egyptologists for two centuries; it just doesn't make sense either as a room or a passageway. The gallery's most striking feature is its height—a soaring twenty-eight feet, which, when compared to the other cramped passageways in the Pyramid, makes it seem even higher. It leads up to the King's Chamber, but if it were only a passageway to a room, why the great height? There are other inexplicable features. The base of the Grand Gallery is eighty-two inches wide—almost seven feet—with stone benches twenty inches wide and two feet high running along both sides of the walls for 150 feet, almost the entire length of the Grand Gallery.

What is the purpose of these benches? They constrict the walking space to forty-two inches. Carved into the benches are slots at regular intervals of one and a half cubits. (A cubit, the primary unit of length in ancient Egypt, was approximately twenty inches.) Again, to what purpose?

The function of the mysterious Grand Gallery has puzzled Egyptologists for decades.

The Gallery is corbelled in seven stages, each stepped inward three inches. At the top the span is forty-one inches, much wider than the corbelled ceilings in earlier pyramids. The only aspect of the Grand Gallery easily explained is the 50 percent upward slope toward the burial chamber. It is an easily measured incline. For each cubit you build up, you go out twice the distance. This is called a 50 percent grade. Such a slope is simple for workmen to follow, but this aspect of the Grand Gallery is about the only one we do understand. The Grand Gallery doesn't make sense as a passageway, but it doesn't make sense as a room, either. It certainly wasn't meant to hold the pharaoh's treasures; the proportions are all wrong for that. What was its function?

As Jean-Pierre worked out the details of his internal ramp theory, he began to think more and more about the Grand Gallery. Then it hit him. Perhaps the function of the Grand Gallery is connected in some way to the next major construction, the King's Chamber. Is there some aspect of the King's Chamber that is unique and would have led to the construction of something like the Grand Gallery? At first glance, the answer was "no." But when he looked again at the details of the King's Chamber, the answer was "yes." Forty-three huge granite beams ranging from thirty to more than sixty tons each were used to build the King's Chamber. They had to be raised more than 140 feet onto the Pyramid, and nothing like that had ever been done in Egypt.

Stone benches with slots in them line the Grand Gallery.

The angle of the Grand Gallery's incline is a 50 percent grade. For each cubit of height you extend 2 cubits horizontally.

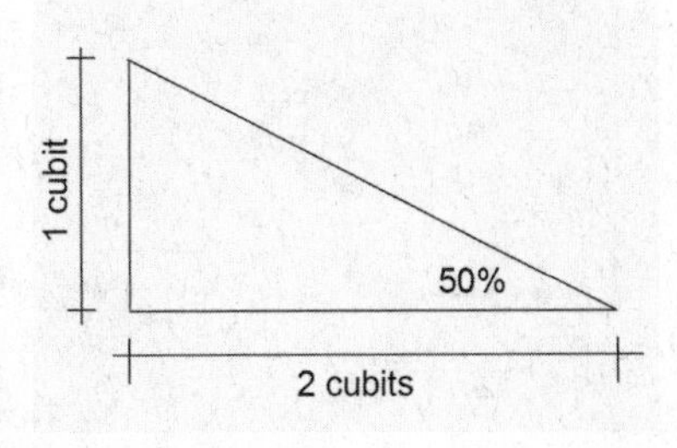

The Egyptians were undeniably skilled at moving heavy stones. Two thousand years after the Great Pyramid was built, they were moving and erecting 250-ton obelisks, but we don't have any ancient records of how they did it. Our best indication of how they moved large stones is painted on a wall of the tomb of Djehuti-hotep, who governed one of the provinces of Egypt about 500 years after the Pyramid was built.

I love this painting. It reminds me of the scene in the 1923 Cecil B. DeMille silent film *The Ten Commandments* where the oppressed Israelites are hauling stones up a ramp. If you look closely you will see that the men are hauling the stones up the unfinished obelisk that lies at an angle in the Aswan granite quarry. With just as much drama as the film, the tomb painting shows a colossal statue of Djehuti-hotep being pulled by 172 men. The statue rests on a sled and a worker standing by the statue's feet pours water or oil in front to lubricate the way. There is even an overseer, standing on the lap of the statue, who is clapping the rhythm to which the men pull. Moving the statue was no easy feat. In the painting we see men carrying large wood beams, perhaps replacement parts in case the sled breaks. The hieroglyphs on the tomb wall tell the story of moving the statue and how everyone was happy to participate. Some divisions of the military were called in, but priests were also hauling on the ropes. Locals pitched in, even the old and infirm. "The aged one among them leaned upon the boy, the strong-armed was with the trembler [palsied]. Their hearts rose, their arms became strong."[37] I think there must have been a similar feeling of camaraderie

Tomb painting of a statue of the nobleman Djehuti-hotep being pulled on a sled by 172 men.

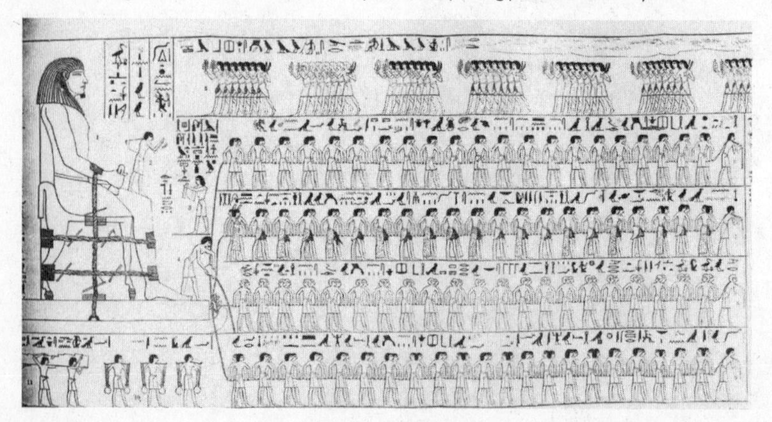

among Hemienu's workers, a feeling they were participating in something great.

The statue was thirteen cubits (twenty-two feet) high and weighed approximately fifty-eight tons, as much as some of the huge granite blocks that had to be raised forty-three meters up the Great Pyramid to be used in the King's Chamber. But there is an important difference. The statue was pulled on level ground; Hemienu had to raise the blocks up a long ramp. The largest block used to build the King's Chamber is sixty-three tons and we can calculate that it would take approximately 630 men to pull a sixty-three-ton block up the Pyramid's ramp. It would have been extremely difficult to coordinate so many workers and difficult to position so many men on the ramp. For centuries Egyptian builders had used manpower to haul heavy blocks, but now that the blocks had to be raised 140 feet, something entirely new had to be invented. As Jean-Pierre created three-dimensional models of the Grand Gallery, his architect's mind concluded that the Gallery wasn't ceremonial, it was functional. As he added more and more details to the computer model, he realized its function. The Grand Gallery was built to raise the huge granite beams.

Hemienu designed a counterweight system to raise the largest blocks, and the heart of this system was the Grand Gallery. Think of the Grand Gallery as a railroad track with the stone benches as the rails on which the car will run. That's why the Grand Gallery is so narrow and that's the purpose of the stone benches. Now imagine one of the huge granite blocks at the bottom of the ramp with ropes attached to the block that run up the ramp and at the other end are attached to the trolley car at the top of the Grand Gallery. The trolley is loaded with four-and-a-half-ton granite blocks that serve as counterweights. When the brake is released, the trolley slides down the Grand Gallery and the large granite beam is pulled up the ramp. It is a bit more complicated than that, but not much. Jean-Pierre began to fill in more and more of the details.

To reduce the friction between the bottom of the car and the bench (remember, the Egyptians didn't use wheels) and ensure that the counterweights move smoothly down the Grand Gallery, rollers made of logs were placed across the width of the Gallery, their ends resting on the benches. The rollers thus form skids, very much like those used to slide boxes of groceries from the delivery trucks down into the basements

When the counterweights were released and slid down the Grand Gallery, they pulled a large granite beam up the external ramp.

of supermarkets. The twenty-eight pairs of slots in the benches (spaced one and a half cubits apart) were used as part of a braking system for the trolley and counterweights, as well as for keeping the logs in line. By inserting wood beams with carved ratchets into the slots, the trolley could be held stationary at almost any point in the gallery.

The full run of the counterweight trolley was seventy-five cubits, about 130 feet, not long enough to pull the beam up the full length of the ramp. Eleven trolley trips were needed to raise each beam from the base to the level of the King's Chamber. Every time the trolley made a run down the Grand Gallery, the beam would move 130 feet up the ramp, and remain there until the trolley was brought to the top and the rope shortened and reattached to the beam for another journey—downward for the trolley and upward for the beam. Hemienu probably realized that there was a way to speed up this whole process. Instead of the entire length of the Gallery being spread with rollers, only half the Gallery had rollers. Imagine that the rollers are at the top of the Gallery, stretching only halfway down, and the counterweight trolley rests upon them at the top. Now we release the brake and the rollers turn as the trolley slides downward. But that's not all. Because the rollers are not fixed and only cover half the Gallery, they slide down the Gallery along with the trolley, so the trolley is actually moving twice as fast as the rollers. The trolley moves along with the rollers—that's one unit of speed—but as the rollers turn the trolley also moves down along the rollers, a second unit of speed. This would have been another ingenious labor-saving device that cut time from the construction of the Pyramid.

The counterweights, by themselves, do not provide enough force to raise the largest beams. The system would have required an additional eighty men pulling, in addition to those reloading the trolley and so on, but still, this is a tremendous savings from the 630 or so men that would have been needed without the counterweight system. Jean-Pierre is the first ever to suggest the Grand Gallery was designed to lift the larger blocks. The idea is certainly possible, and even seems reasonable, but so far it is just theory. Is there any real evidence that the Gallery was used for this purpose?

Along both vertical faces of the Grand Gallery are long, deep gouges that Jean-Pierre believes once held wood railings used to stabilize the trolley as it went up and down. Raising all the large blocks required approximately 800 round-trips of the trolley and it would

TOP: The trolley containing the counterweights moved on log rollers placed across the stone benches in the Grand Gallery. **BOTTOM:** The slots in the stone benches were used to stabilize the rollers.

TOP: Long gouges on the Grand Gallery's walls may have held wooden "tracks" necessary to keep the trolley in line. **BOTTOM:** When the Pyramid was completed the granite counterweights were slid down the ascending passageway, sealing the entrance from tomb robbers. Three are still in place.

A stone cylinder may have hung down from the trolley into the ascending passageway as ballast.

be surprising not to see some physical signs of this. Running the full length of the stone benches are two thin brown lines that look like racing stripes. These could be the remains of grease used to lubricate the trolley. A very close inspection of these lines shows deep scratches, the kind of wear one would expect if the counterweight theory is right.

Further evidence exists that the Grand Gallery was designed as a counterweight system. At the bottom of the ascending passageway you can still see some of the counterweights. When the Pyramid was completed, the granite counterweights were slid down the ascending passage to block the entrance from robbers. The ascending passageway is intentionally two inches narrower at the bottom than at the top so that when the counterweight blocks were slid down they would be stopped by the narrowing and become wedged in place at the bottom. One of these blocks was chiseled away by the caliph Al Mamun's men in the ninth century when they forced their way into the Pyramid in search of treasure, but two others still remain.

There is another bit of evidence for the counterweight system, but it has been hidden for nearly a century. To maintain the stability of the counterweight trolley and keep tension on the rollers, a cylindrical

A 1911 photograph shows a groove for ropes in the large block at the top of the Grand Gallery.

ballast stone was attached to the rollers. This weig
the ascending passageway.

There was a special V-shaped groove for the r
stone, but you can't see it today. When modern tou
Grand Gallery, at the top they climb up two metal
a large block that enables them to continue on to th
The steps are modern additions, but the block itself has also been restored. A photograph taken in 1911 shows it in its broken state with the groove through which the rope holding the ballast stone ran.

If the primary purpose of the Grand Gallery was indeed to raise the huge stones needed for the burial chamber, then this also explains the unusually wide ceiling span at the top of the Grand Gallery. The counterweight system was not only used to raise the huge beams up the ramp, it was also used to raise them into position on top of the burial chamber. For the entire time the giant beams were being moved, the ceiling of the Grand Gallery had to remain open for the ropes to move freely.

The asymmetry of the Grand Gallery also suggests that the ceiling remained open for quite a while. The Egyptians loved symmetry, so anytime it is broken, there's a reason. To the untrained eye the Gallery

The rigging system in the Grand Gallery used to haul the large granite beams up the ramp.

erfectly symmetrical—a tall, narrow hallway with a corbelled ng—but this is not quite the case. The narrow ends—the south and north walls—look similar but were constructed differently. In the lower end, the blocks of the sidewalls cross those of the north wall, as in weaving. To do this, the north and sidewalls had to have been built at the same time. Not so with the upper south end. Here the blocks of the sidewalls do not cross those of the south wall. The sidewalls were built parallel to each other and at some later date the blocks of the south wall were slipped in place against the sidewalls. Why? So the south end could remain open for the ropes of the counterweight system. As the beams of the ceiling were set in place, the blocks forming the south wall were also put in place. When all the giant beams were in place, the counterweight system had exhausted its use, and the Grand Gallery's ceiling was roofed over. This asymmetry between the two ends of the Gallery is just one more indication that the primary function of the Grand Gallery was not as a passageway or chamber. For centuries adventurers and Egyptologists had it wrong. If Jean-Pierre is right, the Grand Gallery was Hemienu's equivalent of the freight elevator.

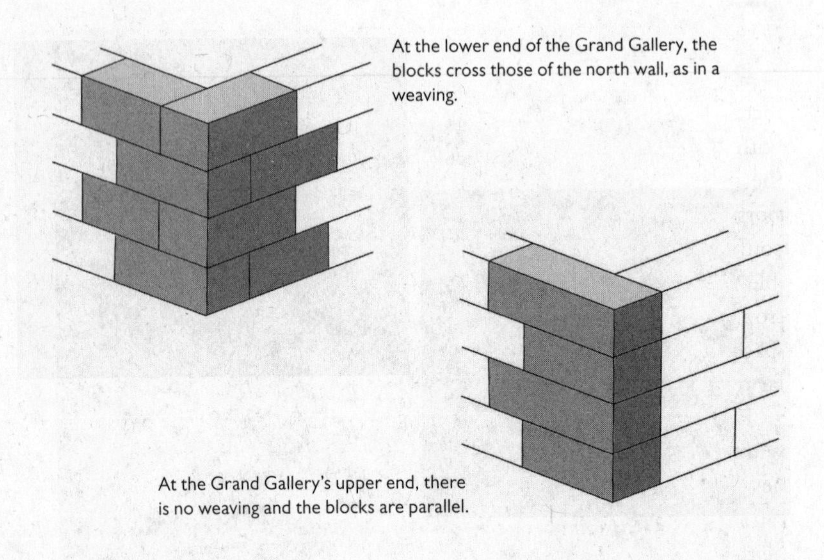

At the lower end of the Grand Gallery, the blocks cross those of the north wall, as in a weaving.

At the Grand Gallery's upper end, there is no weaving and the blocks are parallel.

The Burial Chamber

Giza, Egypt, 2575 B.C. *(Year 14 of the reign of Khufu)*

The primary function of the Grand Gallery was to assist the construction of the King's Chamber, and to an ancient Egyptian the King's Chamber would have been a miracle—a large rectangular room ten by twenty cubits (seventeen by thirty-four feet) with a flat ceiling and thousands of tons of stone above it. How could the ceiling support such weight? In the ancient world, spacious interiors were impossibilities. There were no materials such as steel that could span great distances without breaking, let alone support incredible weight. That's why ancient Egyptian monuments often seem claustrophobic. The Hypostyle Hall at Karnak Temple is a forest of tightly spaced pillars because the columns supported a ceiling and couldn't be far from one another.

Earlier pyramids had used corbelling to construct interior spaces within them, but they narrow toward the top. Hemienu pioneered a new technique with the Queen's Chamber by using huge limestone rafters to distribute the weight above into the body of the Pyramid. But neither of these techniques was used for the King's Chamber.

The flat-ceilinged burial chamber is one of the most outstanding engineering feats of the ancient world, and required incredible plan-

The columns of the Hypostyle Hall at Karnak Temple are closely spaced because they supported a ceiling.

ning. Before the walls were constructed, the king's granite sarcophagus had to be placed on the floor of the room to avoid having to haul it up through narrow passages once the ceiling was built. With the sarcophagus in position, the entire burial chamber was built of huge slabs of Aswan granite finished and set in place so precisely that the edge of a razor blade can't fit between the blocks. As the walls of the burial chamber were constructed, the Pyramid was built up around it to the level of the room's uncovered ceiling. It was now time to move the massive ceiling beams in place. The ceiling is composed of nine twenty-four-foot-long blocks weighing between forty-seven and sixty-three tons each. Once again, the Grand Gallery's counterweight system was used to move huge blocks into position. Khufu now had his flat-ceilinged burial chamber, but the architectural challenges were just beginning. The flat ceiling could not support the weight of the upper courses of the Pyramid that would be built over it, so above the burial chamber a series of four relieving chambers—small granite rooms only four feet high, made of granite beams—were constructed.

The first relieving chamber's ceiling is made up of eight beams weighing between forty-four and sixty-three tons, which were moved

The huge roof beams waiting to be pulled up by the counterweight system.

One of the relieving chambers above the King's Chamber.

into place by the Grand Gallery's counterweight system. Now we are at a height of more than 150 feet and are higher than the Grand Gallery, so the ropes used to haul the beams had to come out of the Grand Gallery's open ceiling. Once again the trolley with counterweights slid down the Grand Gallery and the ceiling blocks were pulled into position. Sometime around the fifteenth year of Khufu's reign, the process was repeated and a second relieving chamber directly above the first was constructed using nine slightly smaller beams weighing between thirty-five and fifty-three tons. The next year the ceiling was put on the third relieving chamber using nine beams weighing between thirty-five and fifty-seven tons. This same year the fourth and final relieving chamber was completed.

The four relieving chambers above the burial chamber have puzzled Egyptologists for years. What was their purpose? We have already said that the flat ceiling of the burial chamber, even though constructed of huge granite slabs, cannot support the weight of the Pyramid above it. But doesn't the same logic apply to the top relieving chamber? With the weight of the Pyramid above, won't its beams crack? The answer is "yes." The relieving chambers don't really solve the problem. That's why, above the fourth relieving chamber, there are rafters—eleven pairs of huge limestone blocks forming an inverted V that take the pressure off the ceiling and direct the forces into the solid mass of the Pyramid. But if that's the case, why not put the rafters directly above the burial chamber and do away with the relieving chambers?

Above the last relieving chamber are huge limestone rafters to take the weight above off the ceiling below.

The answer literally points to the Grand Gallery. If the rafters had been placed directly above the ceiling of the burial chamber, the forces would still have been directed off the ceiling, but those forces would have gone directly into the hollow Grand Gallery, causing it to collapse. Hemienu knew this, so he designed the relieving chambers to raise the rafters above the level of the Grand Gallery. He must have planned this solution from the beginning, knowing he would need all those huge granite blocks and the Grand Gallery to move them into place.

Situating the eleven pairs of giant limestone rafters above the last relieving chamber marked a turning point in the construction of the Pyramid. The burial chamber was now completed and Hemienu must have breathed a giant sigh of relief. He had won the race to build the perfect burial place for his aging pharaoh. All that remained was to

If the limestone rafters had been placed directly above the King's Chamber, the Pyramid's weight would have been directed into the Grand Gallery and it would have collapsed.

build the uppermost part of the Pyramid, a small mass (27 percent of the volume) compared with what had been completed. The upper part contains no internal chambers or passages, so it should have been clear sailing. It wasn't.

Hemienu's Solution

Giza, Egypt, 2573 B.C. (Year 16 in the reign of Khufu)

Once Hemienu completed the King's Chamber, three-quarters of the volume of the Pyramid was in place, but Hemienu was far from done. The last quarter of the Pyramid presented an engineering problem no Egyptian architect had ever encountered before. No one had ever raised blocks 480 feet in the air and the external ramp Hemienu had used up to now wouldn't work for the top blocks. When the King's Chamber was completed, the Pyramid was about 195 feet high. If the ramp were raised to accommodate the increasing height of the Pyramid, the angle of incline would quickly go beyond 8 percent, too steep for men to haul blocks up. If the angle were kept below 8 percent, the ramp would have to be lengthened to more than a mile to reach the top. That would mean the volume of the ramp would increase to about the same as the Great Pyramid itself! From the day he began building the Pyramid, Hemienu knew the time would come when the exterior ramp could not be used, but he left no record of how he solved the problem.

Hemienu must have sketched detailed plans of how he intended to raise the blocks to the top of the Pyramid, but these diagrams have not survived. It isn't surprising that plans drawn forty-five centuries ago have not survived; what is surprising is that *no* construction plans of any kind have survived from ancient Egypt. Construction was the ancient Egyptians' major preoccupation. For 3,000 years they were building big, but no plan survives. We have medical papyri, literary papyri, business transactions, love poetry, but no construction plans, and we don't know why. Some scholars have suggested that building was a trade secret, and perhaps architects didn't want to write down tricks of the trade, but this doesn't seem right. As a temple was being constructed, everyone could see how it was done, especially the workmen. Building wasn't an easy secret to keep.

The answer to why no plans have been found may be in plain sight, carved into the Giza Plateau. The area around the Pyramid is not a flat, smooth surface; there are all kinds of holes gouged into it. Some seem to be postholes that once held poles erected to survey the Pyramid site. The largest holes are huge, gaping 200-foot pits that were cut into the limestone for the boats that would transport the pharaoh to the next world. These boat pits are well known to Egyptologists, and tourists also, but on the east side of the Pyramid, about a hundred feet from the base, is a sixty-foot trench that practically no one knows about. It just might be Hemienu's plan for the Pyramid.

The trench was first recorded in the 1880s by the Englishman William M. Flinders Petrie, the father of modern Egyptology. Petrie stands out as a strange character in a field populated by eccentrics. His frugality on expeditions is legendary. Once a young Egyptologist on one of Petrie's digs noticed that no toilet paper had been included in the expedition's supplies. Afraid of asking Petrie, the young man approached Lady Petrie with the crucial question, only to be told, "Sir Flinders and I use potsherds."[38]

To save on food costs for his expedition, Petrie bought in bulk, which sometimes meant that the archaeologists ate only tinned beef for the entire season. To save on packing the uneaten tins and bringing them back to England, at the end of the season Petrie would bury them on the excavation site and mark the place on the expedition's map. Then at the beginning of the next season the tins would be dug up and the banquet could begin anew.

Petrie's father, a mechanical engineer, had read Piazzi Smyth's book and was interested in the Great Pyramid "for its symbolic interest relating to the higher ideas intentionally embedded therein by its originator," so young Petrie grew up hearing about his father's plan to do a proper survey. For twenty years the father procrastinated, and in the meantime, Flinders became a proficient surveyor and conducted the first careful documentation of Stonehenge. Then in November of 1880, the twenty-six-year-old Flinders Petrie embarked for Egypt, accompanied by crates of scientific instruments.

Soon after landing in Alexandria, Petrie and his crates wended their way to Cairo, where he found Ali Gabri, the same Egyptian assistant Smyth had used. Soon Ali had Petrie ensconced in a comfortable tomb with all the supplies he needed for his survey. Petrie's system of surveying was far more accurate than Smyth's. With the aid of a theodolite and a telescope, Petrie used the surveyor's system of triangulation to take thousand of measurements all over the Giza Plateau. To ensure accuracy, he sometimes took the same measurement a dozen times. After months of working in the scorching Egyptian heat, Petrie calculated that his measurements of the perimeter of the Great Pyramid were accurate to a hundredth of an inch.

Inside the Pyramid, Petrie was just as meticulous. Using a plumb line to determine the vertical, he measured the walls at various heights to detect the tiniest construction errors. Petrie was amazed at the accuracy of the ancient builders. He carefully measured the only object inside burial chamber, the empty sarcophagus made of the same material as the chamber, granite, a material incredibly difficult to work. Petrie concluded that to fashion it so precisely, the Egyptian stonecutters used saws and drills embedded with hard jewels and concluded that the Egyptians had tools better than his. Petrie was dead wrong on this score. There were no precious stones in Egypt. Ancient Egyptian tools were quite primitive, which makes the sarcophagus all the more remarkable. When the Pyramid was built the only metal tools were copper. The sarcophagus was the product of an astounding number of man-hours, cutting, grinding, and pounding away until it was perfect (see Appendix IV).

Petrie didn't just study the Pyramid. He surveyed the entire Giza Plateau, and in the course of his measurements discovered a curious sixty-foot trench cut into the bedrock. He quickly noticed that it is an exact model of the descending and ascending passageways in the Great Pyramid. He calls it "trial passages . . . being a model of the Great Pyramid passages, shortened in length, but of full size in width and height."[39] These trial passages are as finely cut as the passages inside the Great Pyramid. Petrie's nineteenth-century diagram of them shows the angles of both the ascending and descending passages as almost exactly the same as inside the Pyramid, and just like the Great Pyramid, the passages are perfectly aligned north to south. The model is so well crafted that some Egyptologists have suggested it is the abandoned beginnings of a small pyramid.[40] What we are most probably looking at, however, is Hemienu's plan for the Pyramid, not written on papyrus but carved in stone.

There are tremendous advantages to a three-dimensional model over one drawn on paper. The model carved into the Giza Plateau is large enough that you can climb down into it and see how it works. As we have seen, when the Great Pyramid was completed, large granite blocks were slid down the ascending passageway to seal it from tomb robbers, thus the passage narrows very slightly at the bottom, to stop the blocks in just the right position as they slide down. The trial passage narrows in just the same way and perhaps this is where Hemienu tried out his ancient antitheft device, to see if the blocks would slide smoothly down the passage.[41]

Hemienu's ancient model in stone is not unique. A model of a royal tomb carved out of a single limestone block was found during the excavation of the Middle Kingdom pyramids at Dashur.[42] Corridors and chambers are shown in such detail that you can even see how a sliding

Flinders Petrie's nineteenth-century diagram of the plan carved into the Giza Plateau of the Pyramid's passageways.

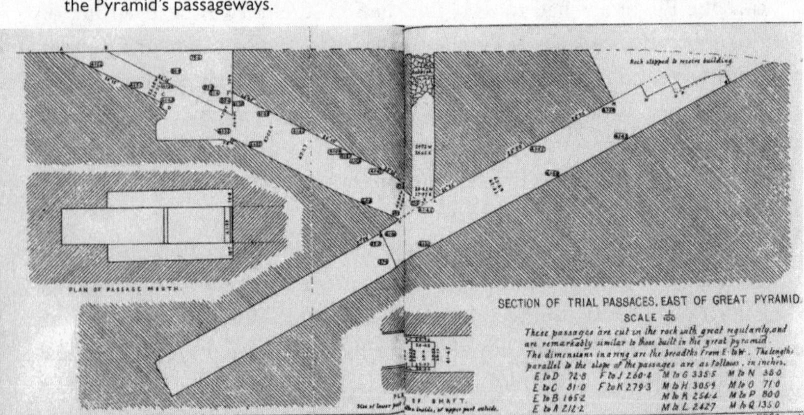

block in front of the burial chamber was intended to seal the king's treasures from robbers. This is clearly an architect's working model used as a guide during the construction of the tomb. Hemienu's model cut into the limestone bedrock probably had a similar purpose.

All this suggests that ancient architects didn't draw elaborate building plans on papyrus, so it is unlikely that we are ever going to find a drawing by Hemienu showing how he intended to raise the two million blocks. In order to figure out Hemienu's secret, we must look for indirect evidence.

The earliest report directly connected to the Great Pyramid is an account by the famous Greek historian Herodotus, who visited Egypt around 450 B.C., two thousand years after Hemienu had completed the Pyramid. Herodotus didn't visit only Egypt, he went all over the known world, and when he returned home he published the first travel book ever, which he titled *Historia*, Greek for "researches," the derivation of our word "history." His account of the pyramids has some glaring inaccuracies and certainly can't be taken as gospel. For example, he says that there is "no block less than thirty feet in length."[43] The truth is, few blocks are more than *five* feet in length. What was he looking at? Still, he has some interesting things to say about how the Pyramid was built.

According to Herodotus, during its early stages of construction, the Pyramid was a stepped pyramid. Then, by means of levers, the remaining blocks were maneuvered up its steps. Herodotus adds that he is not sure if there were levers at each step or if only one lever was used and then carried up step-by-step as a block was raised. Different people told him different methods. The notion of levers has been picked up by several writers who suggest something like a *shadouf* may have been used.

This shadouf was used in ancient Egypt to raise water from the Nile and continued in use through the twentieth century. It is a long pole on a pivot. At one end of the pole is a bucket for water and at the other is a weight, usually a dried ball of mud that makes it easy to lift the bucket of water. There are scenes of shadoufs on ancient tomb wall paintings, and when Napoleon invaded Egypt in 1798, his artists depicted the shadoufs in action. At first it seems reasonable that this water-lifting device could be converted to lifting blocks of stone up the Pyramid, but in reality this idea had insurmountable problems.

Two-ton blocks would require very substantial wood poles and there

Napoleon's artists sketched the *shadoufs* the Egyptians used to raise water from the Nile.

simply was not that much heavy wood in Egypt. Hundreds of shadoufs working constantly would have been needed. Also, there was no room to position the shadoufs. Many levels of the Pyramid are built of relatively small blocks with ledges only about two-feet wide, certainly not enough room for a shadouf. So the idea of levers simply will not work for so many heavy blocks. But Herodotus mentions another possibility.

He says that for ten years the people of Egypt were forced to build the road on which the stones were dragged and comments, "the making of the road was to my thinking a task but a little lighter than the building of the Pyramid." Although it is not certain that Herodotus is talking about a ramp, a later writer of antiquity, Diodorus of Sicily, is quite specific.

Writing about three hundred years after Herodotus, Diodorus says "the construction was effected by means of mounds, since cranes had not yet been invented at that time."[44] This is the earliest mention of the ramp theory and started a controversy that would last for 2,000 years. Did Hemienu use a ramp for the very top of the Pyramid? We are not sure. Diodorus acknowledged that in his time only sand surrounded the Pyramid, with no trace of the mound. Where did this massive construction go?

We know for certain that the ancient Egyptians used ramps for some of their buildings. You can even see one being used if you visit the tomb of an Egyptian named Rechmire, who lived during the 18th Dynasty, more than a thousand years after the Great Pyramid was built. Rechmire was at the top of the food chain—the vizier of Egypt, our equivalent of a prime minister. As the highest official in the land, he was responsible for all sorts of things, from administering justice to overseeing the budgets of various building projects. Naturally, as the vizier, he had a large and beautiful tomb whose walls are decorated with scenes and events from his career. One wall is all hieroglyphs, proclaiming Rechmire's biography and duties. Clearly impressed with his office, one line states, "Let no man judge the vizier in his house." Rechmire wasn't just beyond the law, he *was* the law.[45] But it wasn't all work for Rechmire. One wall of his tomb is covered with beautiful paintings of him hosting a banquet. The ladies, assisted by servant girls, are putting on makeup, people are eating and drinking—clearly Rechmire knew how to entertain.

On the wall opposite the banquet scene, various building projects that Rechmire oversaw are depicted. Several sculptors are completing a sphinx, while another group of workers polish a block of stone. The technique for finishing a block was quite simple. On the wall we see two men stretching a cord on the diagonal of the block. If part of the block isn't smooth and protrudes a bit more than the rest of the block, it will cause a bulge in the string. That's a spot that has to be polished down. Amid all this activity, a building is being erected and to reach the top the workers have constructed a ramp. Interestingly, the ramp is not made of uniform material, such as bricks. It seems to be brick with some rubble as filler.[46] On the bottom of the ramp what looks like a roofing slab is being brought up the ramp to place on top of the walls, which are shown in profile to the right. This ancient representation of a ramp is important, but we can do even better. On the opposite side of the Nile from Rechmire's tomb, at Karnak Temple, is an actual ramp.

A vast complex of temples built over a period of 2,000 years with each pharaoh adding his monument, Karnak covers 300 acres. In front of each temple stood a pylon—a giant gateway, usually carved and painted with scenes of the victorious pharaoh smiting his foreign enemies. The last pylon built at Karnak was constructed in Egypt's decline and was never completed. Even so, it is a huge, impressive mass, nearly 100 feet high and fifteen feet thick, built out of stone blocks. Obviously, some of the blocks had to be raised to the very top of the wall, but here there is no doubt how it was done. The ramp is still in place where 2,000 years ago the workmen put down their tools and never returned. Now eroded and only about thirty feet high, the ramp once reached to the top. From what remains we can see it was constructed of several parallel mud brick walls, and the spaces between the walls were filled in with rubble just like the representation on Rechmire's tomb wall. It is not difficult to use a ramp to build a 100-foot pylon, but

Egyptian tomb painting of a ramp being used in construction.

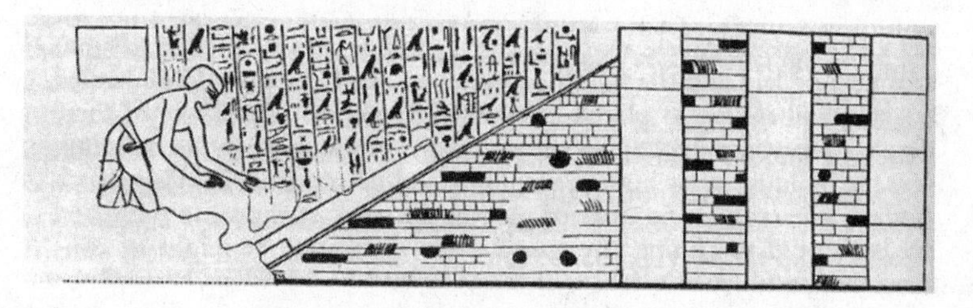

when you consider building a 480-foot pyramid with a ramp, complications arise very quickly.

There are actually two competing theories of how ramps were used to build the Pyramid, and deep down Egyptologists know that both have fatal flaws. The more popular of the two, the external ramp theory, is the easier to imagine. One long ramp on one side of the Pyramid was the highway on which the blocks moved toward the top. As we mentioned earlier, the problem is that there is a limit to how steep the ramp can be and still permit two-and-a-half-ton blocks to be hauled up by workers. Think about how you feel when walking up a steep hill in a city, the kind that takes something out of you, perhaps leaving you winded at the top. Now imagine having to haul not just yourself but a block of stone behind you. That was Hemienu's problem. The hill (ramp) couldn't be very steep, about an 8 percent grade, or else the blocks couldn't be hauled up. In order to have a gentle grade but still reach the top of the 480-foot pyramid, the ramp has to be about a mile long. That's right: as long as fifteen football fields strung together lengthwise. The volume of earth and stone needed for such a huge ramp is just about equal to the volume of the Pyramid itself! So to build the Pyramid using a single external ramp, you don't just build the Pyramid, you also have to build a second structure just as large. That's a lot of man-hours.

Then there is the question of where on the Giza Plateau you can put a mile-long ramp. The plateau is just that, a plateau, and drops precipitously to the desert below on the north side. On both the east and west sides are cemeteries for the nobility that were built as the Pyramid was constructed, so that leaves only one side as a possibility—the south side, the only place a mile-long ramp could have been built. However, no remains of such a large ramp have ever been found. A couple of million tons of stone and earth don't just disappear. Although the single ramp theory is the most popular explanation of how the Great Pyramid was built, it just doesn't seem to work. Time for Plan B, the corkscrew ramp.

To understand the second theory, we have to think in terms of a road that winds up the side of a mountain. If the road went in a straight line, from the bottom up to the top, it would be too steep even for cars. To make the slope gentler, the engineers wrap it around the mountain. It's longer, but a lot less steep. This is exactly what has been sug-

Remains of an ancient ramp used to build a wall at Karnak Temple.

One theory of how blocks were raised to the top of the Great Pyramid is via a ramp corkscrewing up the outside.

gested for the Great Pyramid: that corkscrewing up the Pyramid, like a mountain road, was a ramp on which the blocks were raised. The virtue of this theory is that we don't need a ramp stretching a mile beyond the Pyramid. The theory also explains why the remains of a giant ramp haven't been found—the Pyramid itself was the ramp. It all sounds so good, but Plan B also has a fatal flaw, one that only a builder would notice.

As the Pyramid rose year after year, the angle of incline had to be carefully monitored. With a structure as large as the Great Pyramid, if you are off by an inch at the bottom, by the time you reach the top you could be off by yards and the Pyramid's four edges won't meet at a point. Thus, as the Pyramid rose, surveyors had to make repeated measurements along the edges of the Pyramid to make sure the angle was constant. If, however, there were a ramp corkscrewing around the Pyramid, the sight lines along the edges would have been obscured and Hemienu couldn't have watched the angle as carefully as he needed to. So Hemienu did not use the corkscrew method. Whatever solution Hemienu found, it wasn't what people have proposed. Four thousand five hundred years after Hemienu figured out how to raise the blocks to the top of the Pyramid, a French engineer was closing in on the solution.

First Plans

Paris, 1999

Soon after watching the television show on the Great Pyramid, Henri Houdin sprang into action, putting his new idea that the Pyramid was built from the inside out onto paper. He was an engineer and began by imagining what he would do if he were awarded the contract to build Khufu's Pyramid. For the first time in 4,500 years, a builder was thinking like Hemienu. What materials would he use? How would he bring them to the construction site? How do you move the stones into place high up on the Pyramid? Soon graph paper, pencils, and T-squares were out and he was drawing plans, all the time trying to think like an engineer in ancient Egypt.

Now that he was retired, Henri could devote himself full-time to his new project. As an engineer, he knew that both the single external ramp and the external corkscrew ramp that he had seen on the television show were impractical. There had to be another solution. Then it came to him—build the ramp *inside* the Pyramid! With an internal ramp, the blocks could be brought up the Pyramid without obscuring sight lines, and as the Pyramid grew, so did the ramp. But what would such ramp look like?

Henri Houdin's first drawing of the internal ramp is dated January

4, 1999. It shows a single ramp looking very much like a spiral staircase. In the upper right-hand corner of the plan is the notation "ramp with 8 percent slope." Connected to the top and right sides of the ramp are a series of parallel lines—ventilation ducts so the workers would have fresh air as they hauled the blocks up the internal ramp. But then Henri thought a single ramp would not be sufficient.

By January 24, a second plan had been drawn and it shows four ramps winding through the Pyramid. There are four separate entrances for the four separate ramps, each reaching a different level of the Pyramid. Henri thought he had solved the internal ramp problem. He was far from the solution, but he had taken the first big step forward.

Through the cold Paris winter Henri refined his drawings, trying all the time to think like an Egyptian. It didn't always work. At first he imagined the blocks pulled on wheeled carts. It seems perfectly reasonable, except that the Egyptians used sleds; wheels sink into sand. There were other false starts, but by the end of the summer he was confident enough to try out the theory in public. In October of 1999 Henri Houdin published a brief article on his internal ramp theory in the journal of the French National Society of Engineers and Scientists.[47] He had discussed the article with his architect son and added Jean-Pierre's name as coauthor, but the architect knew there were some serious problems with the plans his father had drawn. Henri's idea that the inside, not the outside, determined how the Pyramid was built was brilliant, and the concept of an internal ramp was a quantum leap forward. Sometimes in science The Big Idea comes from an outsider, someone who hasn't been indoctrinated about what is possible and what isn't. Henri was that outsider. Knowing practically nothing about ancient Egypt, he was free of preconceptions about how the pyramids had been built. The father and son talked about the Great Pyramid every time they met. For Jean-Pierre it was just an interesting avocation his father had taken up, something to keep a retired engineer's mind sharp. Soon he was teaching his seventy-five-year-old father how to use the latest computer software to improve his drawings. But as Jean-Pierre looked over his father's drawings, he knew certain aspects couldn't work.

First, you can't pull heavy blocks of stone up a circular ramp. For efficient pulling you must move in straight lines; you can't be turning all the time. The second problem was that many blocks inside the Pyramid weigh more than sixty tons. To pull a sixty-ton block up an 8 percent

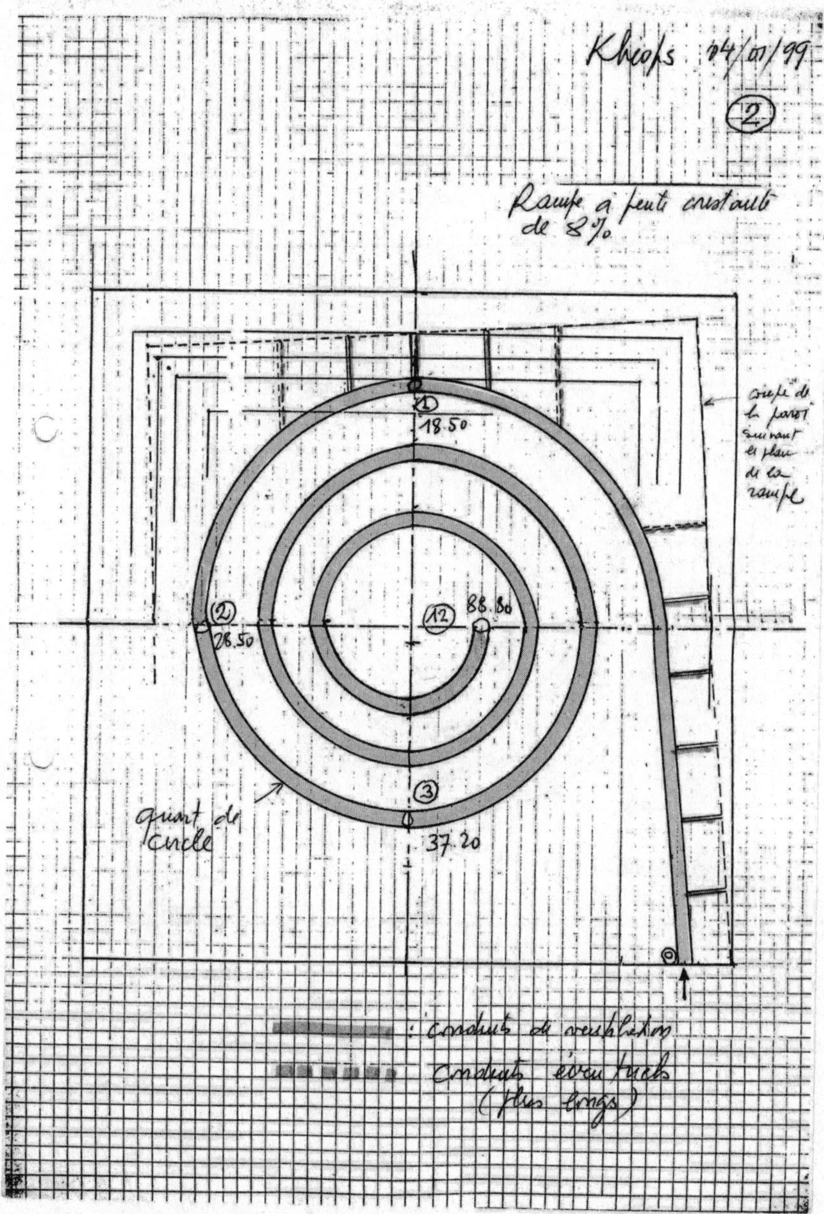

Henri Houdin's first drawing of the internal ramp dated January 4, 1999. Ventilation shafts for the workers are indicated.

Khéops 9/01/99

Galeries multiples

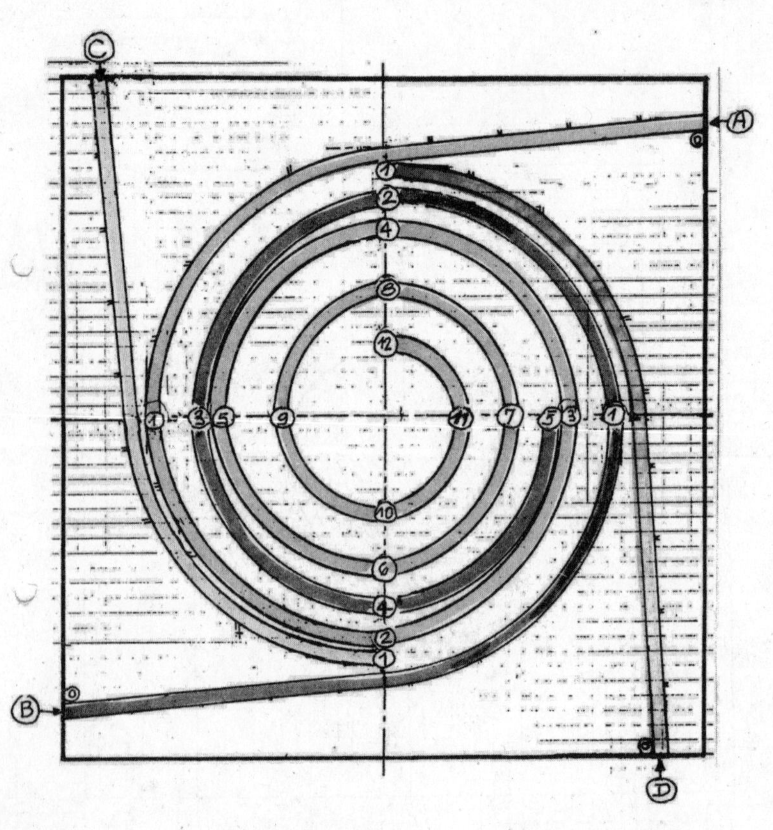

niveau ① 34%
niveau ② 49%
niveau ⑤ 75%

Henri's revised drawing with four internal ramps.

slope would require hundreds of men, and you simply can't get that many men inside an enclosed ramp. Therefore Henri's internal ramp couldn't be circular and the largest blocks couldn't be pulled through any kind of internal ramp. Jean-Pierre was becoming intrigued. This was a problem he would tackle, but unlike his father, he didn't use pencil and paper.

The first step was to learn more about the Pyramid. With his architectural background, Jean-Pierre quickly absorbed the information presented in the standard works on the Pyramid, but soon realized that even the experts were lacking an essential component to solving the mystery. All the plans were drawn two-dimensionally. They all showed the Pyramid's basic components: underground burial chamber, Queen's Chamber, King's Chamber, and Grand Gallery, but only on one plane, a cross section. How were these rooms related to one another in three dimensions? Buildings are designed in three dimensions. All the pyramid experts trying to figure out how it was built were looking at flat, two-dimensional plans. Jean-Pierre was about to become the first person since Hemienu to fully understand the interior of the Great Pyramid in three dimensions.

During his sabbatical year, Jean-Pierre had obtained some of the necessary computer skills he would need to tackle the mystery of the Great Pyramid. Now he went one step further. Using software recently developed for architects, he began to build computer simulations of

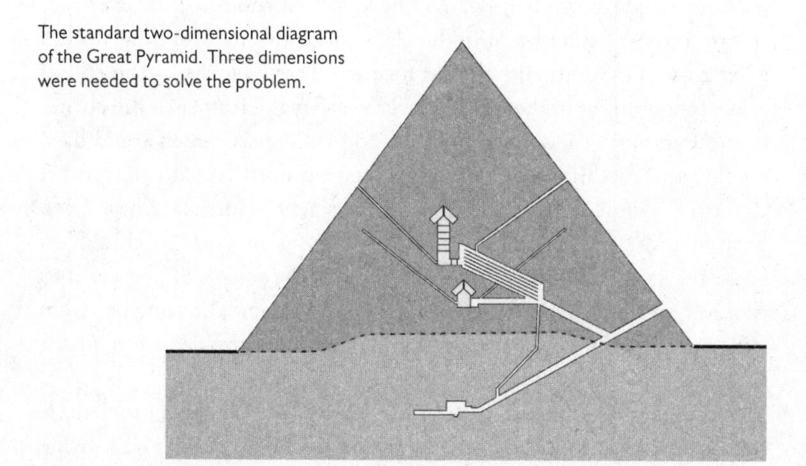

The standard two-dimensional diagram of the Great Pyramid. Three dimensions were needed to solve the problem.

the Pyramid. Even with the new software, re-creating the interior chambers of the Great Pyramid in three dimensions was incredibly time-consuming. Each wall had to be built line by line to the exact proportions of the Pyramid. In his Paris apartment, Jean-Pierre spent first hundreds and then thousands of hours at his computer creating the most detailed three-dimensional renderings of the interior of the Great Pyramid ever. When the models were complete, he could select any internal chamber, rotate it on his computer screen in any direction, see how the chamber related to all the others, and begin to understand the Pyramid's internal secrets. But this was just a start.

The next step was to imagine what Hemienu must have seen as he was building the Pyramid. Jean-Pierre created 3-D representations of the Pyramid for each year of construction. Now he could actually *see* the problems that had to be solved as the Pyramid rose above the Giza Plateau. It was clear that when the Pyramid had reached the height of the King's Chamber, something other than the external ramp had to be used to raise the blocks for the upper part of the Pyramid—but what did the something else look like?

Jean-Pierre had now created more than a thousand computer simulations of the Great Pyramid and an image of what the internal ramp had to look like began to take shape in his mind. He always knew that it had to be straight, not circular, as his father had drawn it, but that still left the problem of turning corners. Moving blocks in a straight line through the interior ramp was relatively easy—eight or so men pulling a block on a sled, but you need lots of room to turn a corner when you are hauling a large block because the men pulling need a place to stand. So, to allow room for turning the blocks, at the end of each straight flight of the ramp, a large space was left open at the corner of the Pyramid. To actually turn the corner, the workmen could have employed something like a shadouf, used up until recent times to lift water from the Nile for irrigation. This ancient crane could have been what Herodotus meant when he said the Egyptians used "machines" to build the Pyramid. The block could be lifted, rotated 90 degrees into position for hauling up the next flight of the ramp. Thus notches were left open at every corner—that would also have provided fresh air for the men working inside the ramp.

Designing the path and size of the ramp was a complex problem. It had to be wide enough for the blocks and men but couldn't be so wide

The internal ramp had corners left open for turning the blocks.

Reconstruction of a possible lifting crane for turning the blocks at the corners.

that its construction created engineering problems. Further, the path of the ramp had to be designed so it didn't intersect with the Pyramid's other passageways and chambers.

Jean-Pierre calculated that the ramp was about five cubits wide (nine feet) with corbelled walls reaching a height of fifteen feet. The corbelling ensured that the weight of the Pyramid above it wouldn't crush the ceiling. At its start near the bottom of the Pyramid, the ramp sloped at about 7 percent and ran straight until it neared the end of one side the Pyramid. Then it made a left turn and continued up the next side at the same 7 percent slope; then another left turn, up the next side and so on up the Pyramid. The Pyramid narrows as its height increases, so the flights of the ramp get shorter and shorter as it corkscrews through the Pyramid. The first straight stretch at the bottom was 575 feet long, but after the internal ramp had made fourteen turns and was higher up in the Pyramid, it would be only 150 feet long. In all, the twists and turns wind through the Pyramid for more than a mile.

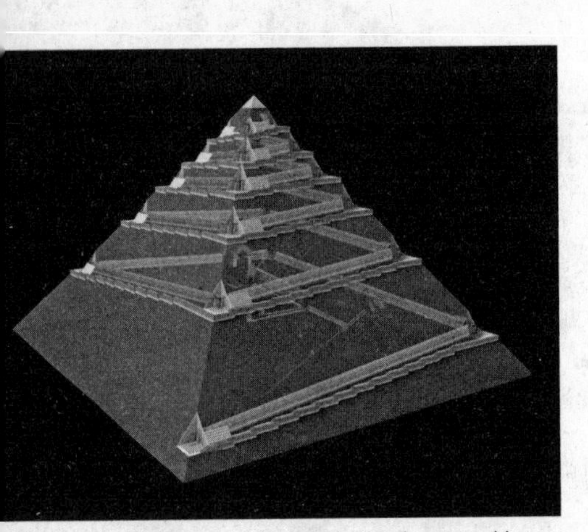

The internal ramp was designed so it would not intersect with any passageways or chambers.

Jean-Pierre now had worked out the basics of his theory of how the Great Pyramid had been built. In a sense three methods had been used. 1) The blocks on the lower levels of the Pyramid were brought up on the short external ramp. 2) The largest granite blocks used for the King's Chamber also came up on the external ramp, but with the help of a counterweight system inside the Grand Gallery. 3) The smaller blocks on the upper levels of the Pyramid were brought up via an internal ramp. The theory was revolutionary, but it seemed to solve all the problems other theories had left unresolved. Still, it was just a theory, and with no physical evidence to support it, it would remain just another theory. The Houdins desperately needed hard evidence to support the theory. When it came, it was from a most unexpected source.

Anomaly Rising

Paris, August 2000

In the summer of 2000 Henri Houdin visited the French company that had conducted the microgravimetric study of the Great Pyramid in 1986, but fourteen years later not everyone who had been on the team was still with the company. Henri met with Pierre Deletie, the company's geologist. When he showed Deletie Jean-Pierre's drawings of what the internal ramp must look like, Deletie was amazed. It looked remarkably like an anomaly his team detected that had remained in the files for fourteen years. He promised to look for the printout and give Henri a copy.

True to his word, Mr. Deletie sent the printout of the anomaly to Henri. Henri was astounded. It was as if someone had photographed their theory. Spiraling up the Pyramid was the internal ramp! Henri excitedly called Jean-Pierre, who rushed over to see what was so amazing. As soon as he saw it, he knew that he had found the big idea he was looking for. The internal ramp was too important to give up. He had to develop the theory a far as it could go, fill in even more details, and most important, see if more evidence could be found to support it.

Seeing how important this was to the Houdins, Deletie suggested they talk with Hui Duong Bui, the man responsible for the calcula-

The first empirical evidence for the internal ramp was this 1986 computer-generated image of low-density areas within the Great Pyramid. They look remarkably like the internal ramp.

tions that produced the image. A week later, Henri and Bui were deep in conversation about the Great Pyramid of Giza. Bui explained that they hadn't totally ignored the anomaly; they just didn't know what to make of it. The team's leading theory was that they were picking up ancient traces of a ramp that once corkscrewed up the outside of the Pyramid. Perhaps somehow all the hauling had compressed the area, making it denser. Deep down they knew this couldn't be right, but it was the best explanation they could come up with at the time. Once Bui heard the details of the internal ramp theory, he was a convert. An internal ramp spiraling upward through the Pyramid is the only thing that could cause such a printout. The internal ramp theory had its first hard evidence.

The Houdins were incredibly encouraged to hear that the French had found evidence for the internal ramp without even knowing it. With this new ammunition, they were more positive than ever that they were right. Jean-Pierre returned to his computer to refine his models, to see if he could figure out more of the details of construction. The carefree days of traveling were over for a while. He wanted to stay in Paris to meet people to discuss and refine the theory. Michelle had seen the obsession coming and quickly realized how big Khufu's Pyramid would be in their lives; she told Jean-Pierre, "If you want to go with Khufu, go straight to the end. You will have to explain everything." They rented a small apartment in Paris in their old neighborhood and for the next year Jean-Pierre worked intensely at his computer, trying to understand exactly how the Grand Gallery fit into the picture with the internal ramp. Often he called his father to say, "You are the engineer, can you do the calculations to figure out how many men are needed to haul such a block of stone, how much manpower was needed for this task?"

By October of 2001, the money from the sale of their apartment was running out. For three years Jean-Pierre and Michelle had lived on that sale, without any income, and now the bank account was dangerously low. Fortunately one of the renters in one of their studio apartments moved out. Without hesitation, they sold it so Jean-Pierre could continue the Pyramid quest. During the next year, he began to complete details of the connection between the internal ramp and the Grand Gallery. But by the end of 2002, the money was running out once again. Another rental apartment became free and was sold. Jean-Pierre was still in the game, at least for a while. As Jean-Pierre refined his theory more and more, he decided to talk to the French team again to see if he had missed anything, or perhaps they had forgotten something.

The Notch

Paris, January 2003

At the beginning of 2003, Jean-Pierre had another meeting with the French Opération Khéops team that surveyed the Pyramid. He described the theory of the internal ramp in detail and mentioned the angle of incline and the notches at the corners that had to be left open so the blocks could turn the corners. One of the team members, Jean-Pierre Baron, remembered something peculiar he had seen when taking measurements up and down the Pyramid. At about 275 feet up the Pyramid, on the northeast corner, was a notch! This was essentially where the Houdins' theory would place the ramp's ninth notch! The French team measured the notch when they were doing their study and mentioned that it was seven feet six inches on each side—a square. In ancient Egyptian terms that translates to five cubits square, a perfect Egyptian measure. The internal ramp theory now had its second empirical confirmation, but there was more to come.

Baron also remembered seeing a desert fox disappear into a hole next to the notch. Why would a fox climb 275 feet up the Pyramid to find a hole? Perhaps it didn't. Perhaps it found an entrance hole to the ramp near the bottom and then made its way up the ramp. It would be interesting to attach a telemetry device to a fox and send it through the hole and track its movements. Would it show the animal moving through the internal ramp?

The internal ramp seemed more and more promising, but there was something else that Jean-Pierre had to explain. Where were the remains of the external ramp used in the early stages of the Pyramid's construction? Although the ramp was used only up to about a third of the Pyramid's height, it was still a huge construction using thousands of tons of materials, but it had vanished without a trace. Somehow the missing external ramp was tied to the construction of the rest of the Pyramid, but how? Then, all at once, it became clear; he knew where the ramp was. He had found the solution. Of course, Hemienu had beaten him to it by 4,500 years.

A notch on the northeast corner of the Pyramid may be a remnant of one of the internal ramp's corners that was left open.

The Internal Ramp

Giza, Egypt, 2570 B.C. (Year 19 of the reign of Khufu)

When Hemienu saw the rafters above the last relieving chamber put in place, he had the Grand Gallery roofed over completely. It was no longer needed; no more huge granite blocks required lifting. At this point the Pyramid was 200 feet high and the exterior ramp that had been used for the first sixteen years could no longer be used. If it were raised to accommodate the increasing height of the Pyramid, the angle of incline would quickly go beyond 8 percent, too steep for men to haul blocks up. It was finally time for the internal ramp to serve its purpose.

From the beginning, Hemienu knew the time would come when the exterior ramp could not be used. For that reason, he constructed it out of rough limestone blocks just like the filler blocks inside of Pyramid, only a bit smaller. Now that the exterior ramp had served its purpose, it was dismantled and the blocks with which it was constructed could be reused to build the rest of the Pyramid. Thousands of years later, archaeologists would puzzle over where the ramp had gone. The answer was in plain view, on top of the Pyramid.

Once again Hemienu's incredible planning was on display, but no one saw it. The top of the Great Pyramid was built from the blocks of the dismantled exterior ramp. The limestone blocks that composed the ramp are slightly smaller than the average block used in the interior of the Pyramid. The blocks had to be smaller because they were going to be hauled through the confined space of the interior ramp. So as the exterior ramp was being taken down, its blocks were moving through the passage up to the top of the Pyramid where they were pushed into position. This explains why no one has found huge piles of debris from the ramp.

As the Pyramid neared completion, Hemienu must have felt both relief and a great sense of satisfaction. Khufu was alive and well, the burial chamber was complete, and the blocks from the external ramp were now moving up through the internal ramp and being placed at the top of the Pyramid. In just a few years the capstone would be in place on the greatest monument ever built. But just as Hemienu was feeling as secure as he could about the project, disaster struck. The burial chamber's roof beams cracked. They were small cracks, but Hemienu had reason to be concerned. Just forty years earlier, the walls of the Bent Pyramid had cracked during construction, and the pyramid had to be abandoned. Hemienu could not afford to abandon the Great Pyramid at this stage; his pharaoh was aging and there was no time to build a second pyramid.

He needed to know what was going on above the ceiling. Were other parts of the Pyramid cracking? At the top of the Grand Gallery he had his stonemasons chisel a small thirty-inch tunnel through the limestone so he could enter the relieving chamber above the ceiling of the King's Chamber. As he crawled through the tunnel he had no idea of what he would find. Had the second relieving chamber come crashing down into the first? He crawled into the pitch-black little chamber and moved his oil lamp around. No disaster.

The floor of the relieving chamber consists of the tops of the ceiling beams and Hemienu could now examine the cracks from both sides of the beams. It didn't look serious, but it could get worse. He had workmen plaster the cracks on the beams. This was not to stabilize the beams; plaster won't do that. Rather, the plaster would tell him if the cracks got larger as the Pyramid was completed and more and more blocks were piled above. Hemienu's technique is similar to what archaeologists do today to tell if monuments are stable. If there is a crack in the wall, something looking like a plaster Band-Aid is placed across the

Hemienu, the pyramid's architect, cut a hole in the top of the Grand Gallery to assess the damage when the Pyramid cracked.

The cracks in the beams of the first relieving chamber are still visible.

Plaster strips on the cracks in the King's Chamber help Egyptologists detect if the cracks get larger.

crack. If the wall moves, the crack enlarges, the plaster cracks, and the archaeologist knows he has a problem. Hemienu's roof beams held, and although the ceilings of two of the relieving chambers above cracked, the plaster didn't move, so he never had to cut a second tunnel going up into the second relieving chamber to check on those beams.

Giza, Egypt, 2569 B.C. (Year 20 in the reign of Khufu)

By Year 20 of Khufu's reign, the Pyramid was nearing completion. It had reached a height of 360 feet and 98 percent of the volume—2,550,000 cubic meters—was in place, but there were still about 130 feet to complete. With the end so near, changes in the internal ramp were necessary. The area on top of the Pyramid was getting quite small. At this point the straightaways of the ramp were only 125 feet long and as they approached the top they decreased to only eighty feet. To avoid running into the corner notch below it, the angle of the ramp had to be increased considerably. It was more difficult to pull the blocks up the steeper incline, but with a small block, weighing only half a ton, this was still possible. However, the internal ramp was nearing the end of its usefulness. By the time it reached the 400-foot level, the angle of incline was 20 percent and almost all the energy of the men hauling would have gone into just raising their own bodies up the ramp.

There is a mechanical trick that the haulers could have used. If they stood on level ground at the notch ahead of the block and pulled, then none of their energy would go toward raising their body weights. This is the kind of technique Hemienu would almost certainly have learned from the previous century of building in stone.

At the top of the Pyramid, the internal ramp's angle of incline had to be increased to avoid hitting the notches at the corners below it.

The workers may have stood on level ground to pull the blocks up, thus saving the energy of having to raise their own weight.

PREVIOUS PAGE:
Along with the three pharaohs' pyramids at Giza are several smaller ones for their queens.

LEFT:
The pharaoh Sneferu built three huge pyramids and developed the techniques for building an internal room in pyramids. His is the first instance of a king's name depicted in a cartouche.

BELOW:
Entrance to the Step Pyramid of Saqqara, the first pyramid in history.

The staircase at the Ptolemaic temple of Dendera may be a descendant of the internal ramp inside the Great Pyramid.

The New Kingdom temple of Medinet Habu still retains some of its original colors.

From earliest times the pharaoh was associated with Horus, the falcon god.

The sphinx has the head of King Kephren and the body of a lion.

The 5th Dynasty pyramids at Abu Sir are much smaller than the pyramids of Giza.

The fallen statue of Ramses the Great at his mortuary temple, the Ramesseum, weighed nearly 1,000 tons.

The Osirion at Abydos was a cenotaph—a false burial—for Osiris, God of the Dead.

The authors planning their Egyptian itinerary. Jean-Pierre Houdin is in black.

Karnak Temple was built by a succession of pharaohs over a period of more than 1,000 years.

The Capstone

The last major engineering task was the positioning of the capstone, basically a small pyramid, on the very top of the Pyramid. The pyramidion, as it is called, was made of fine limestone and probably covered with electrum, an alloy of gold and silver, and might have weighed as much as fifteen tons.[48] We don't know the details of how it was raised, but it certainly would have required advance planning because it is too large to be moved through the cramped internal ramp. After weeks on the computer, Jean-Pierre figured out a possible scenario for the capstone's movements.

During the fourteenth year of Khufu's reign, the pyramidion and the beams and rafters for the King's Chamber were pulled up the exterior ramp to the top of the Pyramid, which was now 150 feet high. For the next five years the capstone moved upward, course by course, as the Pyramid was being built. It could have been suspended in a wooden pyramid-shaped cradle. The ropes holding the capstone could be twisted, much the same way children twist the chains supporting swings. As the ropes twisted, they shortened, and the capstone was raised a bit. Wedges were slid under the capstone, and using wedges, the cradle was raised the same height. Now the process was repeated with the capstone moving slowly upward. This device is called a Spanish winch.[49] When the capstone had been raised to the full height of one course of the Pyramid, normal pyramid blocks were placed under it and the Pyramid course was completed around it. This process continued for five years and more than 100 courses until the capstone was near the top of the Pyramid. At this point working space was very small and special scaffolding was needed to lever and position the pyramidion 480 feet above the Giza Plateau. It was now Year 20 of Khufu's reign and the Pyramid was nearly complete.

During the next few years, the finishing touches are added to the Pyramid. The last fine casing blocks are put in place at the top. The notches left open at the corners of the internal ramp are filled in from the top down with blocks brought through the internal ramp and the ramp is now sealed inside the Pyramid. The last remains of the external ramp and the internal ramp's small exterior access ramp are removed and all traces of Hemienu's construction techniques vanish. Khufu's burial place is ready for eternity. Hemienu has done it. With the exception of the cracks in the King's Chamber, we know of no other close

call endangering the completion of the Pyramid. There was no way that Hemienu could have ever figured out what caused the cracks. The answer to that was buried deep inside the Pyramid.

The hole that Hemienu had his men make so he could enter the relieving chamber would not be discovered until 1764, when an Englishman vacationing in Egypt decided to explore the Pyramid. Nathaniel Davison was visiting Egypt with his friend Edward Montagu, England's former consul to Cairo, and thought it might be entertaining to search the Pyramid for hidden treasure. After unsuccessfully searching the lower portions of the Pyramid, he turned his attention to the mysterious Grand Gallery. At the upper end of the Gallery, he heard echoes of his own voice and deduced that there must be an opening leading to a chamber. Placing a lit candle on the end of a long pole, he probed the darkest recesses of the Gallery and discovered the hole Hemienu had made forty-five centuries earlier. Lashing seven ladders together to reach the opening, he had to clear out more than a foot of bat guano before he could squeeze through the hole. There was, of course, no treasure, only the stifling smell of bats. He left the relieving chamber

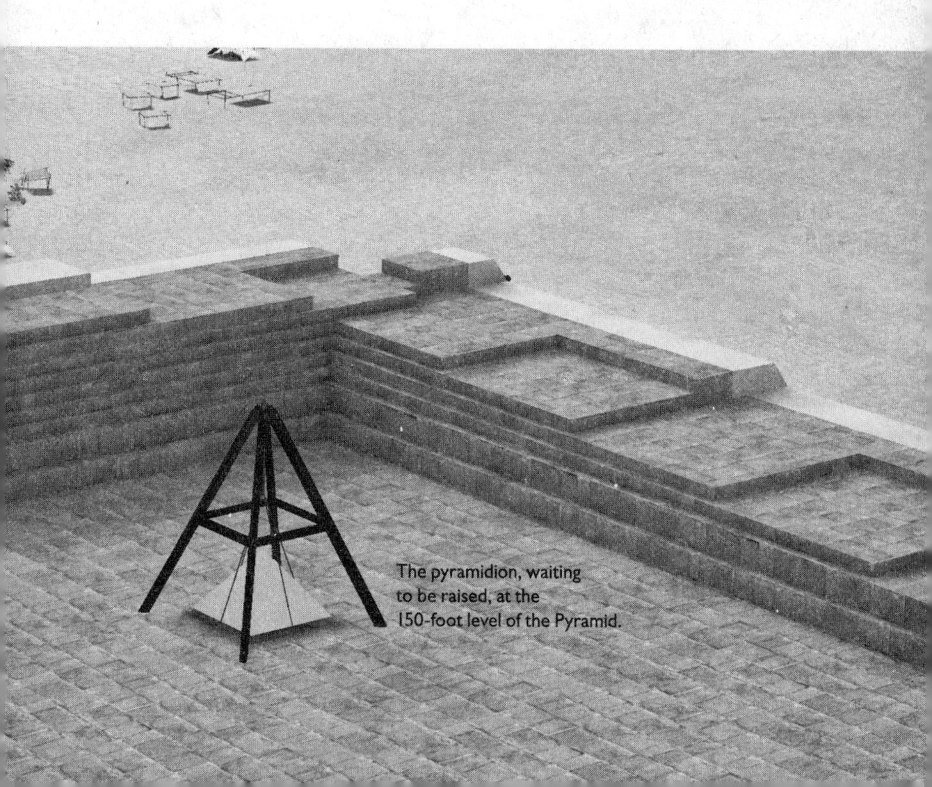

The pyramidion, waiting to be raised, at the 150-foot level of the Pyramid.

The pyramidion enclosed in a "Spanish winch" that lifted it to the top.

with no material rewards, but the chamber is still known today as Davison's Chamber.

During the following decades adventurers would squeeze through the hole, but none went farther than Davison's Chamber until Colonel Howard Vyse appeared on the scene in 1836. Vyse was wealthy enough to hire John Perring, a civil engineer, to assist his investigation.[50] They set up living quarters in a nearby tomb and conducted the most thorough examination of the Pyramid up to that time. Vyse's workmen cleared the Sphinx, which was covered up to its neck in sand, dug up the flooring of the Queen's Chamber, and searched the Pyramid for hidden rooms.[51]

When they investigated Davison's Chamber, they spotted a crack in one of the granite roof beams above them and pushed a three-foot-long reed through the crack, concluding that there must be a chamber above. Soon Vyse's Arab workmen were chipping away at the granite above them. When the granite proved too hard for them, Vyse sent for experienced masons from the Tura quarry across the river. The quarrymen, however, were used to soft limestone and the granite proved too difficult for them, also. Plan C was gunpowder.

A workman named Daoud, sent in to do the dangerous and dirty work, eventually succeeded in blasting a hole through a corner of one of the huge beams forming the ceiling above Davison's Chamber. In the second relieving chamber they found the floor covered with millions of exoskeletons shed by insects that had bred in the chamber, probably soon after it was sealed. Vyse patriotically named the new chamber after Admiral Horatio Nelson, who had sunk Bonaparte's fleet at the battle of the Nile in 1798.

Daoud continued blasting his way upward and discovered the third relieving chamber, which Vyse named after another of Napoleon's defeaters, the Duke of Wellington. Like the others, Wellington's Chamber was empty, so Vyse and his workers continued the difficult process of blasting upward. They were rightly afraid of causing a total collapse of the relieving chamber above them, and when Daoud did set off a charge, he had to watch out for flying granite chips. Reaching the fourth of the relieving chambers, they switched from naming them after British heroes, and named this one after Lady Arbuthnot, who had visited the Pyramid on May 9, 1837, just after the chamber was discovered. At this point Vyse must have been wondering when the succession of relieving chambers would end. Finally, Daoud blasted through to the fifth and last relieving chamber, revealing the huge limestone rafters that Hemienu had set so precisely in place forty-five centuries ago. They named this chamber after England's consul to Cairo, Colonel Campbell.

Vyse didn't find any treasures, but he discovered four relieving chambers not seen since Hemienu constructed them. In addition, he copied the workmen's ancient graffiti on the walls of the chambers and sent it to Dr. Samuel Birch, Keeper of Antiquities at the British Museum and one of the few scholars who could translate ancient Egyptian. Birch was able to read the cartouches containing Khufu's name, conclusively establishing that the Pyramid did, indeed, belong to King Khufu. Vyse was also the first to notice the cracks in the beams of the second relieving chamber, but he had no idea when they cracked or why. That discovery would come much later, when Jean-Pierre acquired a new computer program and a very powerful friend.

The Difficult Years

Just as Jean-Pierre and his father were nearing the solution to how the internal ramp and Grand Gallery worked together in the construction of the Great Pyramid, personal tragedy struck for both. Jean-Pierre's mother was diagnosed with Alzheimer's disease. Over the next few years as Renée Houdin spiraled inevitably downward, Henri was tethered to his house in the countryside taking care of his wife. Project Khufu became his lifeline with the outside world, helping him through the dark years. Every morning and every evening for years, Jean-Pierre would call to give news of the project and to get news about his mother. When her death finally came, Henri had lost thirty pounds and was a broken man, but with Jean-Pierre's encouragement, he began to reconnect with the outside world. He was needed; Project Khufu was in trouble.

Michelle and Jean-Pierre were living in a minuscule studio apartment of only 236 square feet with a bed that folded out of a false closet. Michelle edited her experimental videos at one end of the only table while at the other end Jean-Pierre worked on the Pyramid project. Jean-Pierre assured Michelle, "Don't worry. We will only be living here for a few weeks." (They stayed four years.) They had calculated that with no traveling, they could hold out for nearly a year. The theory was almost complete. Jean-Pierre had concluded that the counterweight system in the Grand Gallery was used to bring the granite ceiling beams up the external ramp and also that the top of the Pyramid was built from the blocks used to construct the external ramp. Many other details were also in place, but without outside help they would soon have to give up the project.

Jean-Pierre knew that he needed the support of the Egyptological community. For him, the United States was a place where new starts were made, new ideas were welcomed, so he took the directory of American Egyptologists and began sending e-mails. One, Jack Josephson, wrote that he should contact an Egyptologist whose name was Bob Brier. That led to dinner at my place and the subsequent meeting with Dieter Arnold. But Egyptological moral support was one thing; Jean-Pierre desperately needed something more substantial.

Jean-Pierre began giving talks to professional engineering groups, both to see how the theory would be received and in the hopes of finding a sponsor. The responses to the theory were overwhelmingly positive. Professional engineers and architects were impressed. The first firm to offer assistance was the Thales Group, a high-tech radar company that was developing a new kind of radar that might be used to detect the internal ramp inside the Great Pyramid. In preparation for a test, they began testing their equipment on Coucy Castle, a thirteenth-century limestone fortress sixty miles north of Paris. If a hidden 4,500-year-old internal ramp was going to be discovered, they wanted to make sure it was with their equipment. Their technical expertise was extremely welcome, but financial support was going to be needed to keep Project Khufu afloat.

Jean-Pierre's brother, Bernard, arranged a meeting with Eiffel, an engineering firm with a pyramid connection—they built the glass pyramid in front of the Louvre in Paris. Jean-Pierre presented the internal ramp theory to a small group of their engineers and executives. It seemed to go very well, but he couldn't be sure. The next day he received an e-mail from one of the senior engineers. "I think you are at the beginning of a discovery far more important than King Tut's tomb." Eiffel was keen to support the project and was transferring 25,000 euros to the project's war chest. A very encouraging first step; the ball was beginning to roll.

Bernard was able to arrange another meeting for Jean-Pierre, this time with the Air France's public relations manager for Africa and the Middle East. After six months of phone tag, the meeting finally took place and Jean-Pierre, laptop in hand, explained his theory over lunch. At the end, Mr. Brousse, the manager simply said, "Welcome on board, dear partner!" Air France couldn't give money but they would fly Jean-Pierre for free anywhere he had to go for Project Khufu. Now he could reduce expenses and wouldn't have to pay for trips to New York and Cairo. Jean-Pierre had finally started visiting the Pyramid!

Soon other sponsors came on board. Everyone wanted to be part of the discovery. French and Egyptian companies contributed cash to a war chest that now held 100,000 euros. Project Khufu was finally on firm ground and the solitary architect in black was no longer alone. With sponsorship came recognition. On May 18, 2005, Jean-Pierre was again giving a lecture in Paris about his work on the Great Pyramid, but not to raise funds. He was receiving the Montgolfier Prize, a reward given by SEIN (Société d'Encouragement pour l'Industrie Nationale), a foundation set up by Napoleon in 1802 to encourage national industry. Jean-Pierre received the medal for his contributions to architecture and beaux-arts. Jean-Pierre's proud father joined him onstage to share the award. When Henri Houdin said, "I have a 5 percent share in the story and Jean-Pierre 95 percent," Jean-Pierre replied, "I would never have had the 95 percent without that first 5 percent." Sponsorship and recognition had finally come. And the greatest boost to Project Khufu was just around the corner.

The Internal Ramp Goes Public

Paris, June 3, 2005

In academia, ideas are constantly being tested, first at conferences and later by publication. This is when you really know if what you are proposing works. Because Jean-Pierre is an architect, not an academic, his ideas didn't take this route. One of the first things I noticed when he gave me some of his writings to read is that there were no references. Normally when you tackle a problem, you survey the literature, read everything on the subject you can find, and list your sources. You don't want to reinvent the wheel. Jean-Pierre just sat down at the computer and started working, which put him at a disadvantage. He had not published the details of his theory nor had he had serious discussions with those qualified to evaluate the theory.[52] Some of his earlier informal talks to engineers had created a buzz about the theory and now his father's professional organization, the Société des Ingénieurs des Arts et Métiers, wanted Jean-Pierre to give a talk. It was the perfect chance to test the waters. The civil engineers would give him needed feedback on the theory but they would be a friendly audience.

The hour lecture went well. There were some questions about details, but no one had any serious objections to the theory, or at least they didn't mention any. In the audience was the former manager of Peugeot/Citroën who had been in charge of digital designing and manufacturing. He understood how heroic Jean-Pierre's solitary effort at digitally reconstructing the Great Pyramid was, and he was equally impressed with the theory itself. He thought it was correct. He was familiar with Dassault Systèmes' software and thought that if Jean-Pierre had access to their programs he could take his simulations to another level. Dassault Group is a multibillion-dollar corporation founded by Marcel Dassault, father of the famous Mirage jet. One of their divisions, Dassault Systèmes, designs 3-D engineering software. Without telling Jean-Pierre, he called Dassault Systèmes and told them about this obsessed genius in a studio apartment in Paris who seemed to have solved the riddle of the Great Pyramid on his computer.

Two weeks after his lecture, Jean-Pierre left Paris for Cairo. As he sat having a drink in the garden of his hotel, his phone rang. Richard Breitner from Dassault Systèmes wanted to talk to him about his theory. At first Jean-Pierre was worried because the cost of the call could be charged to his cell phone, but after a minute talking with Breitner, he knew the conversation would be worth the cost. Dassault Systèmes had a new program called *Passion for Innovation* through which they sponsor a scientist who is thinking outside the box and who does not have access to the usual support systems supplied by universities or corporations. The two men agreed to meet as soon as Jean-Pierre was back in Paris.

Paris, June 30, 2005

On a warm sunny June afternoon, Jean-Pierre had his first meeting with Richard Breitner. A forty-year-old engineer with an intense interest in computer graphics, Breitner was in charge of Dassault corporate systems. From the beginning they knew it would be a perfect marriage. Richard had been interested in ancient Egypt and hieroglyphs since he was a boy and he even dressed in black! With Richard was Mehdi Tayoubi, marketing director for the company. Mehdi was about thirty, a champion of what could be done with 3-D software, but his role would be quite different from Richard's. Richard was a high-level technical person; Mehdi was responsible for getting the word out about Dassault Systèmes' new innovations.

Jean-Pierre did his laptop presentation, but unlike the show in my apartment two years earlier, the audience understood everything Jean-Pierre was talking about. After a few hours, they were hooked. They immediately agreed to work together, the only question was how. Mehdi and Richard felt the Great Pyramid was interesting to everyone and would be the perfect showcase for Dassault Systèmes to strut its stuff. As the three talked they became more and more excited. They would re-create the entire building of the Great Pyramid in 3-D simulation. Not only would the images present the theory, they would be used to validate it. Dassault Systèmes often built entire virtual factories for its clients. The idea was to see if everything worked well together. Do the conveyor belts have room around them so they don't overheat? Are the workspaces large enough to accommodate four mechanics working on an engine? Are the cables in the elevators strong enough to lift heavy freight? These questions can be answered in a virtual factory, before the real one is built. This is how Jean-Pierre, Richard, and Mehdi intended to test the internal ramp theory. They would build the Pyramid, build the internal ramp, have little guys on their computer screen haul the blocks up, and see if it worked. Computer modeling was used all the time to test future constructions; now for the first time they would apply this technology to an ancient monument. Jean-Pierre now had the sponsorship and technical support he needed to take the theory further. He left the meeting elated. It was one of the best days of his life. An organization with seemingly unlimited resources understood and appreciated what he was doing and was going to adopt him. The first thing he did was call his father.

Project Khufu began in earnest in September. Jean-Pierre was sent to Dassault Systèmes' software school for a crash course in CATIA, their flagship 3-D software. As Jean-Pierre was learning new computer skills, teams of engineers from different departments under Richard Breitner's direction began to redraw the Great Pyramid in all its glory. For more than a year the team worked to create the complete virtual Pyramid and workforce. Mehdi, Richard, and Jean-Pierre formed the "Triumvirate," each member with his own part to play. Later, Jean-Pierre would describe it as a "dream team, with no one wanting credit." They were all working to transfer Jean-Pierre's vision to a remarkable 3-D animation. Soon it began to take form. The counterweight system in the Grand Gallery moved up and down on virtual log rollers, thousands of workers toiled in virtual quarries, pulled on virtual ropes, and carried virtual water for other laborers. By midyear 2006, the program was operating. There were no obvious contradictions in Jean-Pierre's theory. Blocks didn't get stuck, ropes didn't snap, and workers didn't collapse. It was a great success, but Mehdi had an even more ambitious project in mind.

Mehdi was creating a new Dassault Systèmes Web site and wanted to put the virtual Pyramid on the site for all to see, but he felt that their scientific re-creations weren't the end. He wanted something more, something like films that people paid to see in theaters. As the team worked on the CATIA presentation, their company bought a small start-up company, Virtools, whose software was perfect for 3-D applications. It would take nearly a year to build the Web site and lots of manpower, but Dassault Systèmes was willing to throw whatever resources were necessary behind Jean-Pierre. As the elaborate simulations appeared on the computer screens everyone on the project became even more excited. Imagine a dozen computer wizards solving the riddle of the Great Pyramid by doing what they love to do, and getting paid for it! The re-creations were perfect for the Web site, but when Mehdi saw just how good they looked he decided to go even further. They would hold an international press conference announcing Jean-Pierre's great discovery.

Richard and Mehdi were used to thinking big and having the resources to achieve their vision. They discussed the idea of having Jean-Pierre onstage interacting with a life-size hologram of Hemienu so the two architects could discuss the building of the Great Pyramid. When the technicians were called in to create the hologram, it soon became clear that it wouldn't work. Some other spectacular means of presenting the theory was needed. Fortune smiled on Project Khufu when Mehdi heard that La Géode, Paris's IMAX theatre, was being equipped with the latest equipment to project stereoscopic animations. They quickly agreed to rent La Géode for the press conference and show a 3-D real-time stereoscopic animation of the Great Pyramid being built. A lot of problems had to be solved to show the presentation on a three-

Men in Black: "The Triumvirate" who created the 3-D real-time simulations of the construction of the Great Pyramid. Jean-Pierre Houdin is in the middle with Mehdi Tayoubi on the left and Richard Breitner on the right.

story-high screen, but they believed they could pull it off. An IMAX film costs millions of dollars to make and takes far longer than the few months they allowed themselves. Even more difficult, they would be using a technique more complex that the usual IMAX. This wouldn't be a film; it would be a digital, real-time, interactive presentation, kind of like a video game on steroids.

As the project neared its conclusion, the team took it to La Géode in the late evenings to try projecting it on the giant screen. Dozens of computer techies, their girlfriends, and Dassault Systèmes employees who had heard about the project filled La Géode for all-night sessions to get the bugs out of the 3-D simulation. As the time neared for the press conference, they figured out how to project it. So much memory was required for the animations that they would have to use six projectors simultaneously. This wouldn't be your ordinary press conference.

As the time of the conference neared, I was asked if I could give a five-minute talk about the theory. Five minutes is about all my French could survive, but I agreed to speak. As the date of the conference neared, my wife Pat and I received two tickets from Air France, and on March 27, 2007, we were winging our way to Paris. Dassault Systèmes had made reservations for us at the Méridien Étoile, a very nice hotel. Things were looking good for the conference.

Paris, March 28, 2007

At six-thirty the next evening, Jean-Pierre met us in our lobby to take us to dinner. He was clearly elated. Finally someone appreciated what he had done. We were joined for dinner by a remarkable assortment of people who were interested in the project and who might help in the future. Two young ladies, one Egyptian and the other Lebanese, staged cultural events in Cairo and worldwide. Very bright, they spoke perfect English, French, and Arabic, and they might help the company and Jean-Pierre increase their image in Egypt. Also present was an engineer from Eiffel who built Sir Norman Foster's famous high bridge, the Millau Viaduct, in southern France. He told us that to raise the bridge's pillars he used methods similar to those the ancient Egyptians used to raise their obelisks. Peter Spry-Leverton, who directed a series of documentaries about Egypt, came from England, trying to figure out if there might be a documentary in the internal ramp theory. Jean-Pierre's father, whose insight started all this, was sitting next to an American engineer, Craig B. Smith, who wrote the book *How the Great Pyramid Was Built* and is a supporter of Jean-Pierre's theory.

The conversation was very lively, if not focused. At eight-thirty we boarded a bus sent by Dassault Systèmes to go to La Géode for a three-hour rehearsal. I wondered why so long. When we arrived, there were about 100 techies and their girlfriends, all twentysomethings. Jean-Pierre came up to me, looking a bit concerned, and asked if it would be okay if I didn't give the speech I was brought over to deliver; the program is running long. I was delighted to bow out. As the rehearsal began, I realized why three hours were needed.

The entire presentation was to be hosted by television personality François de Closets, the Walter Cronkite of France, who would have a dialogue onstage with Jean-Pierre before the film was shown. De Closets was the perfect choice. Eight years earlier, he hosted the television documentary watched by Henri Houdin that started the whole project.

First, a stand-in for the CEO of Dassault Systèmes read his introductory remarks. Their various clients were all mentioned and much was made of the fact that architect Frank Gehry said he couldn't have built the Guggenheim Museum in Bilbao, Spain, without their software. At fifteen minutes, the introductory remarks ran far too long, but that was just the beginning. The CEO stand-in also read a dialogue that was almost an hour long, and we hadn't seen the show yet! Next the bridge builder showed slides of his bridge and ate up another fifteen minutes. Next Craig Smith had his fifteen minutes and spoke about "project management" and how many people would have been needed to build the Pyramid.

When the next speaker stepped up, it became clear why this program was running so long. Everyone had been given fifteen minutes, which doesn't sound like much, but no one had done the addition. Next Richard Breitner gave a talk on the making of the 3-D simulation—the show we haven't seen yet. Richard showed slides of how his company's software enables car manufacturers to simulate car crashes and save money by not having to test-crash real cars. He is passionate about what he does; he wanted to share *everything* with the audience.

By the time we saw the presentation, it was midnight and half the audience was asleep. There was no soundtrack; Jean-Pierre narrated the virtual reality images live and his voice was not strong. He's an architect, not a television host. The presentation was good but it too needed to be shortened. As is so often the case, the technical people had fallen in love with their graphics and not thought much about the story line. There were beautiful, amazing images of blocks being pulled through the internal ramp, wonderful flyovers of the workers' village, and dramatic images of the Grand Gallery being used as a counterweight system to haul up huge blocks of granite. It really was a triumph, but it could have been even better with a bit of editing.

The entire program was about three hours. Jean-Pierre came up to me at the end, very concerned. The press conference was a disaster waiting to be presented on a world stage and everyone knew it. Only radical surgery could save the program. At 1:00 A.M. we all piled on the bus to go back to the hotel. It had been a long night and everyone was quiet. They knew it wasn't working.

The next morning there was a meeting at the Dassault Systèmes offices to discuss the press conference. When I arrived, Jean-Pierre told me that in the longest night of his life, Mehdi had decided to cut the program drastically. The bridge engineer was out, Craig Smith was out, and other cuts were being made. This was encouraging; perhaps disaster could be avoided.

After lunch we walked back to the office for a meeting to discuss how to shorten the program even further. De Closets began by rehearsing his dialogue with Jean-Pierre. He was very good, a real pro—animated, engaging, and mercifully brief. He coached Jean-Pierre to answer his questions succinctly. When Jean-Pierre started talking about "microgravimetrics," De Closets said, "That's a big word. Talk about masses, density . . ." There was hope. Richard Breitner's "Making of the 3-D Animation" speech hadn't been shortened. I suggested we focus on the internal ramp. I lost the battle.

The next morning we all boarded the bus at our hotel for La Géode. Jean-Pierre's brother, Bernard, and their father were on the bus, as were the usual suspects. This was the first time I had met Bernard and he looked nothing like Jean-Pierre. He was very corporate, well manicured, and in an immaculate suit. We arrived a bit early; it was drizzling but lots of people were already there and there was a buzz. We wandered around, had tea and coffee, and, for the first time, met Michelle, Jean-Pierre's wife! We thought she just might be a virtual wife—it was reassuring to see she was real.

The 400-seat theater was packed. Everyone in the audience sensed this was something special, history would be made here today, and they were all excited to be part of it. Dassault Systèmes' CEO began. He was handsome, charming, and it was evident that he was truly interested in innovation. Next we had de Closets: self-assured and concise. He introduced Jean-Pierre and they had their dialogue. It was fine but still a bit too long. Then Richard Breitner came on the stage and we saw the car crash, the cracks, the beams, and everything else. At this point everyone had been listening for more than an hour.

De Closets returned and explained that Jean-Pierre's work was not just a theory, it was a hypothesis that had been tested by the virtual-reality process. Finally we all put on our 3-D glasses and the show began. Jean-Pierre narrated and explained that Fabien would be our pilot. Jean-Pierre asked Fabien to take us up the Pyramid, go down, inside, whatever we wish. The graphics are wonderful; we saw the external ramp, the Grand Gallery being used to haul up the huge ceiling beams, and the internal ramp. We watched a bird fly around the workers' village and it was great fun. At the end there was uplifting music, the bird flew, and we followed it. Lots of applause.

De Closets and Jean-Pierre discussed the possibility of testing the theory. De Closets took questions from the audience but there were very few. Were they exhausted? Confused? Hungry? Then, like the late-night television commercials for the Ginsu steak knives—"But wait, there's more!"—Jean-Pierre wanted to show everyone a thirty-second video that his wife had shot in Cairo from their hotel room window in 2005. It was workmen piling up sandbags to test the structural strength of a foundation for a building they intended to build. They were basically building a pyramid with an internal ramp out of sand bags! Then the program was really over and there was a standing ovation for Jean-Pierre that lasted for a full five minutes. It was very moving. I knew how long and hard Jean-Pierre had worked; this was his day and he was beaming. Over the next few days Jean-Pierre's theory was reported in almost every major newspaper and scientific magazine around the world.[53] The conference had been a smashing success.[54]

tribune

The audience at La Géode in Paris viewing the world premiere of Dassault Systèmes's 3-D animation of the building of the Pyramid.

The Time Machine to Hemienu

The 3-D simulation of the Great Pyramid's construction was a great achievement. It presented Jean-Pierre's internal ramp theory and, to a limited extent, validated it. However, the newly formed trio of Jean-Pierre, Richard, and Mehdi wanted to go even further. They wanted something absolutely certain, "an Egyptological breakthrough that no one could dispute." They decided to tackle the problem of the cracks in the King's Chamber.

In a previous chapter I described how the beams cracked as the Pyramid was being built. Until the trio focused their resources on the cracks, this was far from established. For more than a century Egyptologists were not sure when the beams cracked. They could have cracked centuries after the Pyramid was built, perhaps during an earthquake. Now, using yet another Dassault Systèmes program, SIMULIA, they would act like forensic engineers to determine exactly what caused the cracks. SIMULIA is used by many airplane and carmakers to test the strength of their products: wings and hulls before planes fly, crash tests for cars, and so on. They would treat the Great Pyramid like a construction disaster and "back engineer" what had happened.

Their construction forensics required three steps. First, they needed a geometric modeling, a re-creation in three dimensions, of the Great Pyramid. They had used the plans drawn by the French Opération Khéops team in 1986, which were the most detailed ever produced.[55] But a geometric model isn't enough to simulate physical events. All it can give you is the shape, the geometry of the construction. To re-create an event such as cracking of the beams you also need the physical characteristics of the materials used in the Pyramid. You need the weight of the materials, their elasticity, their texture. For this Jean-Pierre met with François Schlosser from Paris's Bridges and Roads Laboratory, who was able to supply the specific parameters of the limestone and granite used in the Pyramid, which were entered into the computer. The virtual Pyramid was becoming more and more detailed. The last feature needed was functional modeling. Jean-Pierre's theory of how the Pyramid was built involves mechanical systems such as sleds and trolleys running on wooden rollers. The mechanical properties of these systems—friction generated by a block pulled on a sled, compression of a limestone rafter as it is levered into place—all had to be entered into the computer.

Then they built the burial chamber layer by layer. When it was complete, they built the relieving chambers, and finally the uppermost limestone blocks were piled on top of the Pyramid. The engineers expected that as the Pyramid grew the load would be too great for the beams and they would crack. But as they piled more and more blocks on top, the beams held without cracking. Finally the virtual capstone was placed on the Pyramid and still no cracks. One of the computer engineers working on the project telephoned. "Jean-Pierre, we are unable to break these beams; there is no load on the beams, just on the rafters." Of course, this is exactly what Hemienu had planned. The rafters on top were intended to distribute the load through the body of the Pyramid and the computer simulation confirmed that's just what they were doing. The team even simulated a mini earthquake, but still no cracks. And yet the real beams had cracked—and there had to be a reason. Jean-Pierre went back to study the French team's detailed plans of the Pyramid to look for something wrong. He discovered that the rafters had slipped just a tiny bit, just twelve millimeters on the north sides, so that the inverted V-shaped roof pressed on the limestone blocks beneath them, rather than distributing the forces into the body of the Pyramid. This could have transmitted pressure to the beams.

The engineers went back to their simulations, now entering the slight slippage of the rafters. Sure enough they saw considerable pressure on the beams, but no cracks. There had to be something else, so Jean-Pierre went back to the plans. Then he noticed that the south wall of the burial chamber had settled and was three centimeters lower than the north wall. The computer engineers now added the fact that the south wall was slightly lower than the north and ran the simulation again. First they completed the King's Chamber and then went on to build the relieving chambers one at a time. The simulation showed increasing pressure on the beams, but still no cracks. The Pyramid grew course by course on their computer. When it reached a height of about 375 feet, the beams in the King's Chamber cracked. It wasn't just the weight of the Pyramid above that did it; it was a combination of factors. The south wall settled ever so slightly—enough to cause the rafters to shift. This shift transmitted forces downward onto the beams of the ceiling. The granite beams, pinned asymmetrically under the load, had only one way to free the pressure: they couldn't bend, so they broke. From the simulation we now know that the ceiling cracked during the construction of the Pyramid, about three years after the King's Chamber was completed.

The Dassault Systèmes team didn't stop when the ceiling beams cracked. They kept building the Pyramid on the computer, piling more blocks on top. When the Pyramid reached about 420 feet in height there was enough weight above to crack the ceiling beams in the first relieving chamber. At about 450 feet, the ceiling in the second reliev-

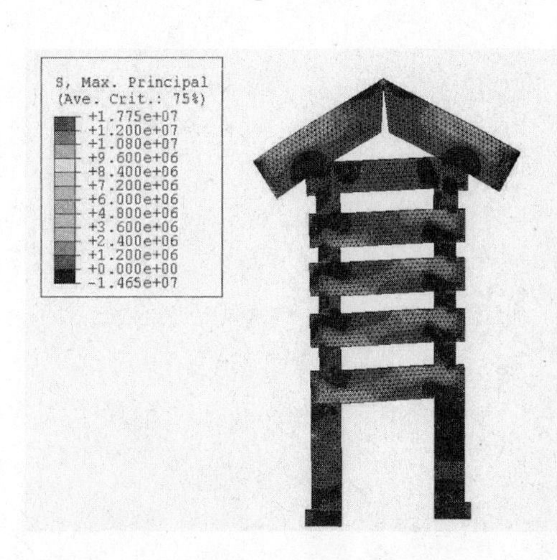

Computer simulation of the cracks in the King's Chamber demonstrated that the Pyramid cracked as it was being built.

ing chamber cracked. At this point there was only another thirty feet or so to the top of the Pyramid, not enough weight to cause more cracks and disaster was averted. The team had conclusively demonstrated not only the cause of the cracks, but when they occurred, the first time such high-powered simulation software has been applied to a building constructed in the distant past.

It was a wonderful discovery—an Egyptological breakthrough that was certain—but something bothered Jean-Pierre. What could cause the south wall to settle soon after the King's Chamber was completed? Why hadn't the other walls settled? With his encyclopedic knowledge of the Pyramid's year-by-year construction, Jean-Pierre came up with the answer. Soon after the completion of the burial chamber, the external ramp was dismantled—the external ramp that had been leaning on the south side of the Pyramid. When this tremendous mass was removed from the south face, the burial chamber's south wall settled just a few centimeters, but enough to have great consequences.

The Search for the Internal Ramp

The Dassault Systèmes press conference thrust Jean-Pierre's theory into the spotlight and opened it up for criticism. Newspaper and magazine reporters were calling pyramid experts for their opinions. Many were noncommittal; they simply didn't have enough details to judge. The good news was that none of the experts found any serious flaws in the theory. The internal ramp was a real possibility, but the details had to be presented somewhere. I published a streamlined account of the theory in *Archaeology* magazine[56] to see what the reaction would be. There was quite a bit of interest, and again none of the experts came up with a serious flaw in the theory. The theory was moving into the mainstream, people were talking about it, referring to it in publications, but it clearly needed some solid evidence.

Much of Jean-Pierre's evidence for the internal ramp had come from computer reconstructions or computer printouts. The notch was the only "real world" support for the theory and that was far from conclusive. Jean-Pierre needed solid, tangible evidence for an internal ramp to take his theory from the realm of the "possible" to the probable. The search for physical evidence began.

In April of 2008, we had a chance to add to the physical evidence for the internal ramp theory. Every year I guide a group of Egyptology enthusiasts through Egypt. Our first day is spent on the Giza Plateau visiting the pyramids, and inevitably ends with a discussion of how the Great Pyramid was built. Thanks to Jean-Pierre, my description of the Pyramid's construction has changed considerably in recent years and now includes a detailed discussion of the internal ramp theory. I point to the notch 275 feet above us and present the possibility that it is the remnant of one of the corners left open so the blocks could make turns up the internal ramp. I tell the group that the French team saw a desert fox disappear into the notch, and also they said that the notch seemed to them to be a perfect square. Someone in the group always says, "Wouldn't it be great to climb up there and see what it really looks like?"

Of course it would, but you can't just climb the Pyramid. It is forbidden. In the old days, when things were much more relaxed, I used to regularly climb to the top of the Great Pyramid on New Year's Eve with my students. Since the 1980s, this has all changed and climbing is carefully regulated. In the 1990s, I was given permission to climb to the top a few times for television documentaries, but I hadn't been up the Pyramid in nearly ten years. Now I didn't want to climb to the top, just to the notch.

When my tour group left Cairo for home, I stayed behind to help *National Geographic* film a documentary about theories of how the Great Pyramid was built. Filming in Egypt involves getting all kinds of permissions and access to closed sites, and I suggested that they request permission for me to climb the Pyramid, not to the top, but just to the notch. Late in the afternoon of April 25, 2008, I began climbing.

There are traditional routes up the Pyramid, where the blocks are not too large and the stone is stable. In some areas the blocks are very big and difficult to get up, and in other areas the stone is crumbly. Unfortunately, the northeast corner, where the notch is, consists of both large and crumbly blocks. When I learned I would be able to climb, I spent several hours looking for the best route up. It was possible, but it would be difficult and slow going. To make matters even more interesting, on the day I was to visit the notch, the little thermometer attached to my bag was reading 111 degrees Fahrenheit. It would be a hot climb.

As I set out to finally see the notch, my pockets contained a tape measure (to confirm the French team's measurement), a digital camera (to take pictures for Jean-Pierre to analyze), and a flashlight (to shine in the crevice where the fox disappeared). As I began slowly picking my way up the first courses of blocks, the footholds were quite narrow and covered with loose rocks, so it was slow going. After about five minutes of very slow progress, the quality of stone got much better and the climb changed from nerve-racking to positively enjoyable. After about fifteen minutes I could see the notch looming just a few courses above me. Thirty seconds later I was standing in the famous notch.

Jean-Pierre and I had often looked at aerial photos of the notch, trying to make out details that might indicate a ramp. We had endless conversations about the French report and what it might mean for the theory. Now that I was actually standing in the notch, I was disappointed. It wasn't what I expected. It was much larger. From what the French had told Jean-Pierre, I had imagined a rather regular seven-foot six-inch square floor. Now that I was there, I couldn't imagine what they were talking about. I was standing on a very irregular platform twice that size.

I didn't have time to ponder the discrepancy between what my eyes were telling me and what the French had reported. It was hot and I couldn't stay on the Pyramid too long, so I quickly ran through the mental list of my three tasks, figuring out what to do first. The obvious order was to photograph first while I had the light, measure second, again with the light, and third shine the flashlight into the crevice, since that didn't require light. As I looked around me, it became obvious that the third task wasn't going to be what I had imagined either. At the back of the notch was a crevice about eighteen inches wide and five feet high. Behind it was an area that looked like a small room—large enough for a hundred desert foxes! I was eager to slip through the crevice, but decided to photograph the notch first before the light faded.

With my digital camera I took the photos I thought Jean-Pierre would want, trying to show the masonry, the irregular floor, and the position of the crevice. The problem was that I couldn't step back far enough to give him a real perspective. I really needed to be hovering in space beyond the notch so I could get it all in one photo. Detail shots would have to do. As soon as the photos were taken, I was on my hands and knees measuring.

The notch was so irregular, with blocks jutting out in all directions, that it wasn't clear which dimensions to measure. I pretended it was a

The notch, high on the Pyramid, may be the remnant of a corner left open to turn blocks as they moved up the internal ramp.

rectangle and took four measurements. It was quick and dirty, but it was all I had time for. It was time to go through the crevice.

I was greeted by graffiti painted in black on one of the blocks. This wasn't virgin territory. Still, it was interesting, very interesting. I was standing inside a fairly large space. Was this the seven-foot six-inch square the French had talked about? No, it was too large and also irregular. It was as if a cave had been formed by stacking up large blocks on top of each other, forming a roof more than eight feet high. It certainly wasn't the internal ramp, but what was such a large space doing so high up on the Pyramid?

A crevice in the back of the notch led to a large open space inside the pyramid.

I quickly photographed "the cave." The crevice let in enough sunlight that I was sure the photos would come out. As I was measuring the cave for Jean-Pierre, I noticed a second crevice at floor level inside the cave. It was formed by two blocks not being set right next to each other, leaving a space of about twelve inches. But the little crevice extended well beyond the first two blocks, forming a small tunnel that ran for at least fifteen feet. I couldn't squeeze inside, so I took a few photos and started my descent.

Going down is a lot easier than going up. It's the difference between hauling yourself up three- and four-foot blocks of stone and letting yourself down those blocks. With a relaxed trip downward, I started to evaluate the notch and how it fit in with Jean-Pierre's theory. I was a bit disappointed. The fantasy was that I would find a small crevice at floor level, shine my flashlight through, and see what looked like a ramp. Jean-Pierre is proven right, gets to take his walk up the ramp, and is acclaimed by the world as a genius. Fade to black. I knew this was unlikely, but that was the ideal scenario.

A smaller crevice led to a small "tunnel."

On a more realistic level, I had expected to find a more finished, regular area, suggesting it had been constructed for a purpose. To my untrained eye I couldn't see much evidence of this. It just looked like blocks piled on top of blocks in the course of the Pyramid's construction. On the other hand, the notch was quite large, large enough to hold a crane. And what was that large space behind the crevice? Was it just a cavity left in the course of building the Pyramid to save a few blocks of stone? Are there other undetected cavities like it throughout the Pyramid? Do the notch and the space have a common cause? It seemed unlikely that the space behind the notch was just a coincidence. These were my questions as I lowered myself down the Pyramid. Near the bottom I once again had to concentrate on what I was doing. The ledges were getting narrower and the footing uncertain.

As I slowly picked my route down I could see Jean-Pierre waiting for me fifty feet below. Soon I was answering his questions as best I could. While I was a bit disappointed, or at least confused, by what I had found, he seemed elated. It all seemed to fit in with his theory. I was dying for a cold bottle of water and left a bit precipitously. We agreed we would download my pictures onto his computer at the hotel so he could do the analysis.

At the hotel that night we downloaded the photos from my digital camera onto Jean-Pierre's laptop. As the images popped up on the screen, Jean-Pierre pointed out details that my nonarchitect's brain had missed. "Doesn't that look like rough corbelling?" "See that block? It had to have been pushed into position from behind." It all made good internal ramp sense to Jean-Pierre—even the roughness of the notch and the cave behind it. But he didn't want to say more. He was already planning a computer model of the notch and cave that would reveal more about how the cave had been constructed and the kinds of blocks that were now missing in the notch. It would take months before he could say anything for certain, but the search for evidence for the internal ramp was off to a good start.

In the quest for evidence, Jean-Pierre began studying hundreds of photos of the Pyramid, looking for clues to its construction. As he looked at one of the Sphinx from the 1880s with the Great Pyramid in the background, he couldn't believe what he saw—the internal ramp! At least that's what it looks like. Going across the Pyramid is a straight line at approximately an 8 percent slope. This, of course, doesn't make any sense. The ramp should be about ten feet inside the surface of the Pyramid; there is no reason to expect faint traces of it on the outside, although these could be from the parallel gangway down which the haulers descended. This is not the only photo that shows this phenomenon. A more recent photo also shows very faint, ghostly traces of what looks like two courses of the ramp. This kind of evidence is intriguing, but isn't very strong. What is really needed is an actual internal ramp, not faint traces of one.

The closest thing to an ancient Egyptian internal ramp is not in a pyramid, but in a temple at Dendera, 500 miles south of Giza. Built 2,000 years after the Great Pyramid, the temple was constructed when Greeks were ruling Egypt, but the same building technique of placing blocks of stone upon blocks of stone was still in use. One of the most beautiful temples in all of Egypt, Dendera was dedicated to the goddess Hathor, mistress of love and music. Each column in the temple is surmounted by the head of the goddess, a beautiful woman with the

A recent photo also shows traces of what could be the internal ramp.

ears of a cow. (Cows in ancient Egypt were associated with nurturing, mother's milk, and security.)

Egyptian temples were vast, dark enclosed structures. The common person was never supposed to enter into the sacred precinct of the god; that was the realm of the priests. Each day, within the temple's many rooms, priests made offerings of food and drink to Hathor. Once a year, on New Year's Day, there was a solemn procession to the roof of the temple, where shaven-headed priests greeted the sun and gave thanks for its bounty. The route those priests took is still inside the temple, and it is an internal ramp. Unlike Hemienu's internal ramp, this one wasn't used for construction. This ramp was for processions and is narrower than the one Hemienu built inside the Great Pyramid, but it shows that the vocabulary of ancient Egyptian architecture included the internal ramp, even 2,000 years later than the Great Pyramid. Carved on both walls are depictions of priests carrying boxes containing the linens, perfumes, and other offerings they will present to Re, the sun god. Fortunately, the processional ramp at Dendera is not the only one

The Temple of Dendera was dedicated to Hathor, goddess of love and music.

Internal ramp used by priests of Dendera to ascend to the temple roof for New Year rituals.

in Egypt. There is another, much closer to the Great Pyramid, both in space and time.

When Jean-Pierre and I had our meeting with Dieter Arnold at the Metropolitan Museum of Art, just before we left, Dieter casually mentioned that in an earlier excavation at Abu Gurob they found what looked like an internal ramp, but no one knew what to make of it. Abu Gurob is just a few miles from the Great Pyramid and its main feature is a sun temple built by King Ni-Userre of the 5th Dynasty. The 4th Dynasty pyramids of Giza marked the end of huge pyramids. Egypt would never again build anything as large as the Great Pyramid. Later pharaohs built smaller pyramids, and in the 5th Dynasty it was the fashion to construct a small pyramid and also a temple dedicated to the sun. Ni-Userre's sun temple was built about 100 years after the Great Pyramid and provides physical evidence for internal ramps.

The sun temple was basically a very large base on which an obelisk was erected. Obelisks, reaching upward to the sky, were associated with the sun and were made from a single piece of granite. (America's Washington Monument is based on the Egyptian obelisk, but it is built of thousands of blocks of stone.) Ni-Userre's temple was excavated almost exactly 100 years ago by the renowned German Egyptologist Ludwig Borchardt. He published a diagram of the sun temple and it shows what appears to be an internal ramp.[57] If the drawing is accurate, this would

Ludwig Borchardt's diagram of an internal ramp inside a sun temple at Abu Gurob.

Remains of a 4,000-year-old internal ramp at Abu Gurob.

indeed be an important bit of evidence—an internal ramp constructed soon after the construction of the Great Pyramid.

The site of Abu Gurob is reached through a lovely grove of palm trees. Where the palm trees end, the desert begins, and there on a hill is the sun temple. My first glance was disappointing. The sun temple is terribly ruined and clearly Borchardt's diagram was his reconstruction of what once was there, not what he saw. But sure enough, when I walked around a bit I could see that there had been an internal ramp and it was still recognizable. The ceiling was gone, but you can still walk up the path that priests had once taken to make their offerings, turn right, and continue up toward the top. Little more than a century after the Great Pyramid had been built, an unnamed Egyptian architect had paid homage to Hemienu's creation.

What's Next?

Finding the small, battered ramp inside the ruined temple was encouraging, but not the kind of evidence we really needed. In a perfect world, the best evidence would be to find the ramp inside the Pyramid itself. If we were going to do this, we knew it would have to be done nondestructively. There could be no moving of blocks of stone on the Pyramid. Our best bet was thermal photography. Thermal photography detects heat differentials. If the ramp really is inside the Pyramid, then the air inside the ramp would be cooler than the stones surrounding it. With a very good thermal camera, we might be able to do a kind of X-ray of the Pyramid and photograph the cool ramp inside. Thermal photography has been used before in archaeology to determine the internal structure of ancient buildings. We just needed permission to try it on the Great Pyramid.

To work on the antiquities of Egypt, you need formal permission from the Supreme Council of Antiquities. You must submit a description of what you want to do, the résumés of the team members, and the time period you will need to complete the project. The team members must have both the skills and the credentials to work on the project. It all sounds relatively simple, but soon we found out it wasn't so easy.

There are two different kinds of permissions—excavations and surveys. Excavations involve moving earth and sand and is the more difficult of the two permissions to obtain. Surveys move nothing and usually consist of mapping a monument such as a tomb or temple and copying the inscriptions on its walls. Our project was a survey; we didn't want to move any blocks, drill any holes, or move any sand. We just wanted to photograph the Pyramid with a thermal camera or a similar device to detect whether there really was a mile-long ramp inside.

When we wrote our proposal we knew that Jean-Pierre couldn't be the project director. The Supreme Council of Antiquities requires a university affiliation for project directors and Jean-Pierre was an architect, not a university professor. To ensure that our project would be approved we asked Dieter Arnold to serve as director. Well respected and the author of the most authoritative book on building in ancient Egypt, he had excavated pyramids for thirty years. Our proposal explained exactly what we wanted to do and what we hoped to find. We had everything necessary for the project in place, listed the companies supplying all the equipment, and explained that the entire project could be completed in less than a week, and we emphasized the nondestructive nature of our study. With high hopes we sent the proposal to Dr. Zahi Hawass, the secretary-general of the Supreme Council of Antiquities.

Hawass is the gray-haired, enthusiastic Egyptian Egyptologist who is in practically every television documentary ever produced about ancient Egypt. Nothing is approved without him, and for years I have worked with him on good terms. I was optimistic about our approval, but it was not to be. We were rejected because Dieter Arnold already had one concession to excavate, at Dashur, and it was one per customer. The Supreme Council wanted to make sure that the director was properly attending to his project; and no one could be in two places at the same time. This concern went back to the early days of Egyptian archaeology when the Antiquities Service was under the control of the French and its director, August Mariette, conducted as many as a dozen excavations at the same time! Who knows what antiquities were stolen by his workmen when he was off supervising other excavations hundreds of miles away.

We had Plan B. Earlier, Jean-Pierre had discussed the internal ramp theory with Dr. Rainer Stadelmann, a member of the German Archaeological Institute who had also written extensively about pyramids. Like Arnold, he too thought it was an "interesting" possibility. We had thought that if for any reason Dieter Arnold didn't want to be the project director, then we would ask Stadelmann. But he too had his own ongoing excavation in Egypt, so we were never able to ask him. We were getting desperate and moved on to Plan C.

I was a university professor and wasn't excavating. Could I be project director? The truth was that the reason I wasn't excavating is that I have no idea how to excavate. Dirt archaeology is a very different skill from mummy studies. I haven't been trained to lay down grids, survey an excavation site, dig trial trenches, and conserve what I might find. Still, the project was only a survey, and we had other experts on the team. It was worth a try.

Dr. Hawass's assistant e-mailed that this plan wouldn't work either. While Zahi had high praise for my work, he felt my expertise was "the wrong flavor." I was disappointed, but understood the position. Why should a mummy person be supervising a pyramid project? We were running out of plans when we came up with another possibility. Why couldn't Zahi himself be project director? Before becoming secretary-general, Dr. Hawass had been director of the Giza Plateau. He is an expert on pyramids, loved to be around them, and thought our theory was interesting, though he never said he believed it. If we were right about the ramp, this project would get the attention of the world and television documentaries were inevitable. Zahi is excellent on television and could also be the spokesman for the project. Of course, this was too perfect to be true. Zahi declined; this was not his project and he didn't want to do it; he was busy working on his own research. It may also be that Hawass was being polite and didn't believe the man in black's theory. Perhaps he needs more evidence.

It may well be that before we receive permission we have to accumulate even more evidence, build an even stronger case for the internal ramp. Jean-Pierre's analysis of the notch and the "room" behind it may add something. Help may also come from new technology on the horizon. A simple infrared camera enabling you to snap a picture of a building and then see the rooms inside may be on the market soon. Such a camera could help add considerably to the evidence.

As we continue to gather evidence, I am reminded of Scully and Mulder in the old *X-Files* TV series. We just need a bit more evidence to prove the case before the authorities will listen. Our advantage over the *X-Files* is that the end is in sight. Either the ramp is inside the Pyramid or it isn't. The truth is out there and I am certain we will know it soon.

Appendices

Appendix I: The Search for Imhotep

Almost certainly, Imhotep would have been rewarded by Zoser with a great tomb, but in spite of extensive searches by Egyptologists, that tomb has not yet been found. There is considerable evidence that he was buried at Saqqara, near the pharaoh he served so well. We know from ancient writings that two thousand years after his death, when he was worshipped as the god of healing, Imhotep's tomb was a pilgrimage site, like Lourdes in France, visited by the sick in search of miraculous cures. Pilgrims came from all over Egypt—and when they finally reached Saqqara they were greeted by vendors selling mummified ibises. The ibis was sacred to Imhotep, so the idea was that if you bought one of these birds and left it as an offering at Imhotep's tomb, he would be pleased and cure you.

The mummified ibis trade was big business and thousands of these birds were raised in captivity, sacrificed, mummified, wrapped, and then placed in clay pots to be sold to the pilgrims. Because the mummies were wrapped and in pots, the hopeful traveler was buying a "pig in a poke." He had no idea what was inside the offering he had just purchased, and was often cheated. When we study mummies we rarely unwrap them—X-rays and CAT scans are nondestructive and quite revealing. Often, beautiful wrappings conceal just a bundle of rags and a few random animal bones, but no ibis. Hor, the priest in charge of the animal offerings at Saqqara, complained of the fraud. He wanted "a god in every pot!" Regardless of whether the pots contained real or fake mummies, they provided clues that sent Egyptologists searching for decades for the tomb of the architect of the first pyramid.

In 1966, Walter B. Emery, a well-known British archaeologist, began his search for Imhotep. Spotting a trail of thousands of broken potsherds at Saqqara, Emery thought they might have been broken by pilgrims along their route to Imhotep's tomb. During his excavations he found several important tombs of the 3rd Dynasty, the right period for Imhotep. In one tomb he even found 500 pots containing ibises, but still no Imhotep. Emery's next discovery was the entrance to the now-famous animal-mummy galleries. The galleries run for miles beneath the sands of Saqqara, and carved into the walls of the tunnels are niches, each one holding a clay pot. The first gallery Emery excavated held tens of thousands of mummified ibises, convincing him that he was on the trail of Imhotep. Many were elaborately wrapped with appliqué decorations sewn onto the outer wrappings. Pressing on, Emery found a gallery of mummified baboons, each wrapped and placed in a coffin that rested inside a niche carved into the soft limestone. Both the ibis and baboon were forms of the god Toth, who was associated with Imhotep.

Emery excavated for six seasons, discovering miles of tunnels containing more than a million mummified animals. Within the rubble of some galleries were plaster models of various human body parts—hands, legs, feet—undoubtedly left by pilgrims with afflictions of those members, hoping to be healed. In the 1969–70 season, he discovered his first written confirmation connecting Imhotep with the animal galleries. In one gallery Emery discovered an ink inscription that read: "May Imhotep, the great son of Ptah the great god and the good god who rests here, give life to Petenfertem . . ." All indications were that Emery was nearing his goal. He hoped that if he followed the galleries to their end, he would find Imhotep's tomb, but it was not an easy excavation. Some galleries were totally blocked with thousands of mummified birds that had been stacked up when wall space for niches had run out. Once the roof collapsed as they excavated and had to be shored up. In the end, Emery suffered a fatal stroke while working at Saqqara and never found Imhotep. For decades no one resumed Emery's work, but currently several excavation teams at Saqqara are hoping to find the architect of the first pyramid in history.

A Polish expedition working near the Step Pyramid has found blue-green ceramic tiles similar to those used in Zoser's burial chamber, and the team is hoping that Imhotep may have used these tiles for his tomb. In 2008, a Scottish team using ground penetrating radar at Saqqara found a very large tomb beneath the sand. The tomb has not yet been excavated, but they say it is large enough to be Imhotep's. The search for Imhotep's tomb may be nearing an end.

Appendix II: The Lost Pyramid

One might wonder how a pyramid can be lost, but that's what happened just a few hundred yards from the Step Pyramid. In 1951, Egyptian Egyptologist Zakariah Goneim discovered the remains of a pyramid that was never completed.[58] Unlike Zoser's, which had six steps, it was apparently intended to have seven. Almost totally covered in sand, its excavation took several years and yielded much information about the early days of pyramid construction. One graffito on the pyramid's enclosure wall reads "Imhotep," so it is quite possible that Imhotep outlived Zoser and designed a second pyramid. During the 1953–54 season, the entrance was found on the north side, with the door still sealed, indicating that although the pyramid was unfinished, the pharaoh evidently had been buried inside. There was tremendous excitement; an intact royal burial of the Old Kingdom had been found. To make it even better, fragmentary inscriptions indicated the king was Horus-Sekhem-Khet, a little-known pharaoh who was about to roar back into history.

Goneim opened the door and found a descending corridor framed by an arch—so much for the Romans inventing the arch! At the bottom of the corridor, huge limestone blocks and rubble still blocked the entrance to the burial chamber. Carefully removing the debris, Goneim found gold jewelry on the corridor floor: twenty-one gold bracelets, 388 hollow beads, and the remains of a wooden magical wand covered in gold. Why had such a treasure been left on the floor? Perhaps to appease tomb robbers, in the hope that after clearing the corridor they would find the jewelry, be satisfied, and leave. It wasn't necessary: they never got that far, and the door to the burial chamber was still intact when Goneim reached it. It must have occurred to Goneim that his discovery could be the next Tutankhamen. If there was gold jewelry on the floor outside the burial chamber, what was inside? Various officials and news media were told of the great discovery, and on the day of the opening, the corridor was packed with reporters.

The door was carefully taken down and Goneim peered in. Everyone was shocked. The burial chamber was virtually empty! There were no treasures as in Tutankhamen's tomb—no furniture, no boxes with clothing, no statues or vases, just a translucent alabaster sarcophagus. Once the disappointment subsided, everyone's attention turned to the sarcophagus. It was still sealed and on the lid were plant remains, perhaps incense used in a burial ritual. The sarcophagus had obviously not been touched since it was placed beneath the pyramid more than 4,500 years ago, so once again spirits rose. The intact mummy of an Old Kingdom pharaoh was still a fabulous discovery.

It took several hours of hard work to open the sarcophagus, as its five-ton sliding side panel had to be lifted. Again Goneim peered in; again disappointment. The sarcophagus was empty; it had never been used. Where was Horus-Sekhem-Khet? The pyramid was a decoy, intended to throw thieves off the trail of the pharaoh's true burial. If this is correct, it may explain why the pyramid was never plundered. Robbers frequently obtained inside information from tomb builders. If they did in this case, they would have known the unfinished pyramid of Horus-Sekhem-Khet was a dud, not worthy of robbing.

Appendix III: The Case of the Missing Queen

One of the great treasures in Cairo's Egyptian Museum is the funerary furniture and jewelry of Queen Hetepheres. Elegant furniture with beautiful gold bands of hieroglyphs are the main attraction, but there are also spectacular silver bracelets inlaid with turquoise butterflies, whose discovery came about under unique circumstances. Found in 1925, when Tutankhamen's tomb was being excavated, Hetepheres's tomb was another intact royal tomb with (hopefully) the same potential for treasures.

Unlike Howard Carter's discovery, which was the result of hard work and knowing what to look for, Hetepheres's tomb was found by pure luck. A photographer for the Harvard–Boston Museum team excavating at Giza was photographing the site when one of the legs of his tripod seemed to go through the bedrock on which it rested. Careful examination revealed that the tripod leg had actually gone through ancient plaster that was covering a deep shaft. As the rubble filling the shaft was removed, it became clear that there would be a tomb at the bottom. Thirty feet down, the excavators uncovered a sealed wall; behind it would hopefully be something rivaling Tutankhamen's treasures.

When the wall was taken down, a small room was revealed with the remains of ancient furniture, an alabaster sarcophagus, and a few other objects. Not quite Tutankhamen, but still some wonderful things. The inscriptions on the furniture showed that this was the burial site of Queen Hetepheres, the wife of Sneferu and the mother of Khufu, builder of the Great Pyramid. It seems as if Khufu wanted his mother buried close to him, so he had her tomb dug near the base of his Pyramid. The one piece that didn't fit was the size of the tomb; it was rather modest for such an important queen. Then things became even more puzzling.

Now all the excavators had to do was remove the lid of the alabaster sarcophagus in the tomb and they would become the first people in more than four thousand years to gaze on the face of Queen Hetepheres. First the furniture had to be removed to create enough space to work on the sarcophagus. This was extremely difficult as the wood had deteriorated and everything had to be photographed and mapped in place first so that if it crumbled to dust it could be recreated in modern materials. Finally the time to remove the sarcophagus lid arrived and an imposing group of officials was invited to the opening.

One by one the august visitors were roped into an armchair and lowered into the tomb. The master of ceremonies was George Andrew Reisner, field director of the expedition. Reisner had asked the expedition's artist, Joseph Lindon Smith, to be present. He later published an account of the surprising events of that day.

> Wheeler and Dunham were at either side of the sarcophagus, to operate two short projectors, which were to serve as handles for lifting the lid. Fitted under the handles was a frame of wooden beams resting on the jack screws. Reisner sat on a small box, and I was next to him, kneeling, and closest to the sarcophagus. In a breathless silence, the lid began to be lifted. When it was sufficiently raised for me to peer inside, I saw to my dismay that the queen was not there—the sarcophagus was Empty! Turning to Reisner, I said in a voice louder than I had intended, "George, she's a dud!" Whereupon the minister of Public Works asked, "What is a Dud?" Reisner rose from his box and said, "Gentlemen, I regret Queen Hetepheres is not receiving," and added, "Mrs. Reisner will serve refreshments at the camp."[59]

To make things even more puzzling, one of the objects in the tomb was a beautiful chest with four compartments, carved out of a single block of alabaster. It contained the internal organs of Queen Hetepheres that had been removed at the time of mummification. The absence of the queen's mummy led to all kinds of speculation. Reisner's theory was that the queen's original tomb was at Dashur, near her husband Sneferu's pyramid. Unfortunately her tomb was partially robbed during the reign of her son, Khufu. By the time officials discovered that the tomb had been plundered, the body was missing. (Tomb robbers frequently removed the body to unwrap it later in safety, hoping to find jewelry.) Rather than upset the pharaoh with news that his mother's body was missing, the officials resealed the sarcophagus and told the king that although some objects had been looted from the tomb, her body was intact. To prevent further robbery of the tomb, Khufu reburied what he thought was his mother's body, along with what remained of her funerary equipment, near his Pyramid on the Giza Plateau.

Reisner's theory is highly speculative and was perhaps influenced by his addiction to mystery novels. He read hundreds and hundreds of them, which were later donated to Harvard's Widener Library. Each one has Reisner's evaluation on the front endpaper. He graded them like students' papers—many have B+ and the really bad got a C. No tomb for Hetepheres has ever been found at Dashur, so we really don't know if there was a robbery and reburial. It could be that like her husband, Sneferu, Hetepheres had a southern and northern burial and her second tomb containing her mummy is still to be found.

Appendix IV: Making Khufu's Sarcophagus

Lying in a granite quarry at Aswan is a 3,000-year-old obelisk that holds the answer to what tools were used to fashion Khufu's sarcophagus. The obelisk cracked while it was being quarried and was abandoned, still attached to the quarry on two sides. Weighing in at around 1,000 tons—as much as two jumbo jets—the obelisk may be the largest single block of stone ever quarried. Because it is unfinished, we can clearly see the techniques used to free it from the quarry.

There are no chisel marks on the obelisk, but on the sides where it is free from the quarry we can see depressions as if a huge ice-cream scoop had removed balls of granite to shape the obelisk. What tool could have done this? Nothing as sophisticated as Petrie imagined. Throughout the quarry are hundreds of black stones the size of cantaloupes. These are the tools that pounded the obelisk out of the quarry, and also shaped the sarcophagus in the King's Chamber that so impressed Petrie. The black rocks are dolerite, which is even harder than granite. They were repeatedly dropped on the edges of the obelisk, causing the depressions. This is a labor-intensive process, involving thousands and thousands of man-hours to repeatedly drop the pounders to chip away the granite a tiny bit at a time; a task possibly done by prisoners sentenced to hard labor in the quarry. Petrie didn't realize that unlimited man-hours can take the place of fine tools. The techniques that were used to quarry the obelisk were basically the same as those needed to free the block from which Khufu' s sarcophagus was fashioned.

In a remarkable series of experiments, Denys Stocks, an expert in ancient stoneworking technology, re-created the tools used by Hemienu's workers and demonstrated how Khufu's sarcophagus was fashioned.[60] Once the rough block of granite was pounded out of the quarry, the ends were cut with giant copper saws much like old-fashioned two-man saws. The copper saws, however, were simple rectangular sheets of copper with no teeth. Sand, used as an abrasive, was placed between the blade and the future sarcophagus and two men would move the saw back and forth, wearing a groove into the granite block. It was a slow, laborious process, but eventually the end of the block was sawn off. Hollowing out the block was more complicated.

Ancient Egyptian tomb paintings show craftsmen using bow drills—something very much like the bow from a bow and arrow. When the bow was moved back and forth horizontally, the string rotated a tubular copper bit. With sand as an abrasive, a circular hole was slowly worn into the granite. Stocks's experiments showed that to drill just six centimeters into granite by this method took sixty man-hours. More than 100 such holes were drilled into the granite and the cores that were left were knocked out with chisels. By doing rather than theorizing, Stocks could calculate the man-hours needed to create Khufu's sarcophagus—28,000 man-hours![61] This is a figure I always keep in mind when I think about all the granite beams and blocks used for the burial chamber in the Great Pyramid.

Appendix V: The Pyramid's Angle

Before a single block was put in place, the angle of the pyramid had to be determined. The Egyptians did not think in terms of degrees; they never divided the circle into 360 degrees as we do. Rather, their unit for angles was called the *seked.* Their unit of length was the cubit, roughly the distance from elbow to the end of the middle finger. This was, of course, standardized. Their measuring rod, the cubit stick, was divided into seven palms, and each palm into four fingers, similar to the way we divide a yardstick into feet and inches. When Egyptian architects were considering angles, they thought in terms of how many cubits you built outward for each cubit of height. For example, if the pyramid rises one cubit and you build out one cubit, you will have a 45-degree angle.

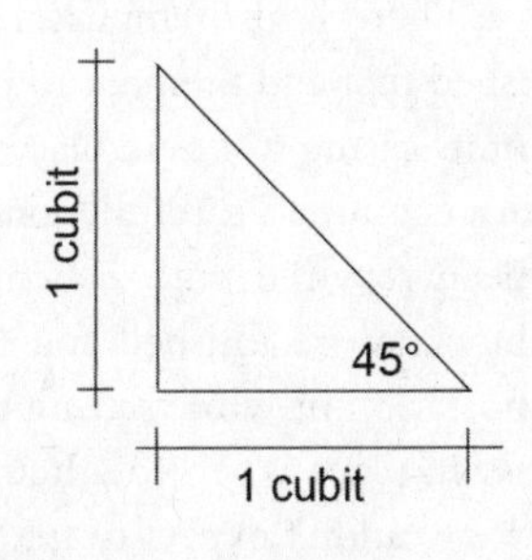

The ratio doesn't have to be in terms of cubits; we could also think in terms of palms. So if you built upward one cubit but out ten palms, you would have a seked of ten, which is the same as our 35-degree angle.

All pyramids may look pretty much alike to the layman, but they are not. At Meidum, the first attempt at the true pyramid, the exterior angle is 52 degrees. The Bent Pyramid begins at 54 degrees, but toward the top it changes to 43 degrees. The Red Pyramid is 43 degrees and the Great Pyramid is 52 degrees. These pyramids all had white casing stones that were crucial to ensuring that the pyramids' angles were constant throughout construction.

The thousands of casing stones needed for the Great Pyramid would have been finished at the quarry with their 52-degree angles before transportation to the site. When the angle is carved on the rough block, the weight is reduced by several hundred pounds. By completing the blocks in the quarry rather than shipping them in rough form to the construction site, you save the shipping and hauling of thousands of tons of what will eventually be rubble.

Herodotus was told that the Pyramid was completed from the top down—that the casing stones were put on the surface of the Pyramid while still in their rectangular shape and then, when all were in place, the blocks were given their 52-degree angle by carving from top to bottom. The economic reason mentioned above gives one argument against Herodotus's account, but there are other arguments. First, where would the workers stand when carving from the top down? There isn't enough room at the top. Second, as a pyramid rises, the angles at the corners must be repeatedly checked to make certain that the four sides will meet perfectly at the top to create a point. This requires that finished blocks be in place as the pyramid grows. Third, if all the external blocks were finished at the end of the project, this would add several years to the construction time. Thus, for all these reasons, it makes sense to complete each facing block in the quarry before it is placed on the pyramid.

There is also empirical evidence to show that the blocks were finished first and then put in place. The Bent Pyramid at Dashur has more of its casing stones in place than any other pyramid. When these blocks are examined carefully, one sees numerous blocks with chips that have been repaired with matching limestone plugs. This suggests that the blocks were finished and some damage occurred either during transportation or while setting the blocks in position. If the blocks had been set in place when rough and then finished in situ from top to bottom, we wouldn't expect to find many chips and repairs.

Endnotes

1. Isler, *Sticks, Stones & Shadows*, pp. 215–217, and Arnold, *Building in Egypt*, p. 100. The problems with the single ramp theory are sometimes ignored. See for example Romer, *The Great Pyramid*, pp. 204–205, where he merely asserts "there was probably a single central ramp built up higher and longer that rested on the Pyramid's south face . . ." There is never any discussion of just how large or long such a ramp would have to be.
2. Burton, *The Book of the Thousand Nights and a Night*, p. 1675. The story of Al Mamun's entering the Great Pyramid is tale 398. The idea that the pyramid contained metal that would not rust and glass that would not break is an early example of the tradition that the ancient Egyptians were an advanced civilization, much of whose wisdom and technology have been lost.
3. Two eight-inch rectangular channels, one beginning in the south wall and one in the north, continue nearly 200 feet through the core of the Pyramid. Why? It has been suggested that the shafts provided ventilation for the workmen inside the chamber, but such a small opening is clearly inadequate for air circulation. Others suggest that they were used for observing stars to assist in the construction of the

Pyramid. This is impossible. The shafts are not straight; they begin horizontally and then angle upward into the body of the Pyramid. Still another possibility is that they were a very early intercom system that conducted sound, enabling the workmen to communicate from different parts of the Pyramid. It is also possible that the shafts had a purely religious function, permitting the soul of the deceased pharaoh to come and go through the Pyramid.

In 1992, in an attempt to solve the mystery of the air shafts, a small robot with a miniature video camera was designed and constructed by Rudolph Gantenbrink, a German engineer. Named Wepwawet, after the jackal god who led the deceased to the next world and was called "the opener of the way," the robot was basically a miniature tank, complete with treads. The idea was to send the 17-centimeter robot through the 19-centimeter shafts to discover where they went. On its first voyage of discovery, it became lodged in the shaft in the south wall. The enormous weight deflected by the chamber's rafters had caused some settling of the blocks forming the shaft and in some places the shaft was only 16.5 centimeters, just too small for Wepawet. In its first attempt, the robot only penetrated 12 meters.

In the northern shaft, the camera revealed a long metal rod that had been abandoned by Waynman Dixon, a nineteenth-century explorer who first discovered the shafts by chiseling into the wall. In 1993 Gantenbrink returned to Egypt with a smaller, redesigned Wepawet. On March 22, 1993, the robot slowly made its way up the southern shaft and after about 185 feet encountered a slab of limestone completely blocking the passageway. On the slab are two copper fittings that look like handles. What's behind the block? We don't know, and the robots' exploration of the shafts in the second burial chamber didn't solve the question of the purpose of the air shafts.

4. Exodus 1:12.
5. Genesis 37:2–50:26. The Joseph story of course never mentions building pyramids as granaries. There are several Egyptian representations of granaries and none is even remotely pyramidal. The Metropolitan Museum of Art in New York has a three-dimensional model of a granary found in the tomb of a noblemen named Meketre. It is a rectangular room.

6. Greaves, *Pyramidographia.*
7. Smyth, *The Great Pyramid*, p. ix. Smyth's great opus went through many editions, even though his theories were discarded by the scientific community, and is still in print today more than a century after it first appeared.
8. Smyth, page 35.
9. Ostrander and Schroeder, *Psychic Discoveries.* The authors visited Soviet parapsychologists and reported their claims that dull razor blades were sharpened when placed inside a cardboard pyramid. When I was a research fellow at the Institute for Parapsychology in Durham, NC, several of us tested the Soviet claim with poor results. Still, the book started the "Pyramid Power" fad.
10. Scamuzzi, *Egyptian Art.*
11. Brier, "The Use of Natron in Human Mummification."
12. Lauer, *Saqqara.* The excavation and reconstruction of the Step Pyramid was the work of the French architect Jean-Philippe Lauer, who worked at the site for more than fifty years.
13. Mendelssohn, *Riddle of the Pyramids.*
14. Another theory is that the pharaoh died before the pyramid was completed and was thus buried elsewhere. This theory relies on the identification of the owner of the pyramid as King Huni, father of Sneferu, but there is little evidence for this attribution.
15. The ancient Greeks had a reverence for the even more ancient Egyptians. Herodotus proudly states that the Greeks got their gods from the Egyptians and also learned how to build in stone from the Egyptians. Plato's *Dialogues* present frequent references to the Egyptians' skills and wisdom. In the 1990s, Martin Bernal's book *Black Athena* continued this tradition and attempted to show that much of western civilization derived from ancient Egypt. This book was highly controversial and was criticized by the majority of scholars.
16. Gillings, *Mathematics in the Time of the Pharaohs.*
17. Lichtheim, "Three Tales of Wonder."
18. Jenkins, *Boat Beneath the Pyramid.*
19. Lipke, *Royal Ship of Cheops.*
20. Manetho, *Aegyptiaca.*
21. Because of the high water table in the area of what was ancient Memphis, the ancient city has sunk into the ground and almost

completely disappeared. Excavations in the 1890s and early 1900s were hindered considerably by water.

22. Krauss, "The Length of Sneferu's Reign."
23. Romer, *The Great Pyramid*, pp. 74–75. It is difficult to estimate the number of workers, and estimates vary considerably.
24. Lehner, in *Giza Reports*, gives a good overview of the excavation of the workers' village. He also publishes a newsletter, *Aeragram*, that gives progress reports of each season.
25. Naville, *The Temple of Deir El Bahr*, Plates CLIII–CLIV.
26. Landstrom, *Ships of the Pharaohs*, provides a good survey of the kinds of boats used in ancient Egypt.
27. Romer, *The Great Pyramid*, p. 169.
28. Petrie, *Researches in Sinai*.
29. Romer, *The Great Pyramid*, p. 158.
30. Lucas, *Ancient Egyptian Materials*, pp. 74–79.
31. Lehner, *The Complete Pyramids*, p. 212.
32. Spence, "Ancient Egyptian Chronology."
33. Goidin and Dormion, "Architectural Analysis."
34. Bui, "First Results of Structural Analysis."
35. Tonouchi, "Non-Destructive Pyramid Investigations."
36. Eissa, "Application of Mossbauer and X-ray Fluorescence."
37. Newberry, *El Bersheh, Part I*, pp. 16–26 and Pl. XII.
38. For an excellent biography of Petrie, full of wonderful anecdotes, see *Flinders Petrie* by Margaret Drower.
39. Petrie, *Pyramids and Temples*, pp. 15–16.
40. Lehner, *Pyramid Tomb of Hetep-here*, pp. 45–50. For a brief account of the discovery of Queen Hetepheres's tomb, see Appendix III.
41. Hawass, "Pyramids and Temples of Egypt," pp. 107–111.
42. Arnold, *Building in Egypt*, p. 9.
43. Herodotus, *History*, p. 427.
44. Diodorus of Sicily, *Library of History, Vol. I*, p. 217.
45. James, *Pharaoh's People*, p. 64.
46. Clarke, *Ancient Egyptian Construction*, fig. 86.
47. Houdin and Houdin, "Construction de la Pyramide de Khéops" pp. 76–83.
48. Hawass, *Pyramids: Treasures, Mysteries, and New Discoveries*, pp. 176–179.
49. No Spanish winch has ever been found, but it is certainly a simple

enough device that the Egyptians could have used it. Wood from such machines would have been reused in later projects.

50. Perring, *The Pyramids of Gizeh.*
51. Vyse, *Operations Carried On.*
52. Actually, Jean-Pierre Houdin and his father published an early version of his theory in book form in 2003 (Houdin and Houdin, *La Pyramide de Khéops*). The book went largely unnoticed, partly because it was so technical and also because the publisher was extremely small and didn't have the means to distribute it. A later version of the theory was published in 2006 (J.-P. Houdin, *Khufu*) in an edition to be sold only in the bookstore in the Egyptian Museum in Cairo. This book too is extremely technical, has mostly diagrams and little text, and could have benefited with professional editing. It too, understandably, went unnoticed.
53. Kiner, *Révélations sur Khéops*, pp. 44–59, and Brier, "How the Pyramids Were Built" pp. 22–27, for example.
54. Dassault Systèmes, *Khéops Révélé.*
55. Dormion, *Pyramide of Cheops: Architecture des Appartements.* The plans are on twelve separate sheets, each covering a different area of the Great Pyramid.
56. Brier, *How the Pyramids Were Built.*
57. Borchardt, "Der Bau," fig. 20.
58. Goneim, *The Buried Pyramid.*
59. J. L. Smith, *Tombs, Temples, & Ancient Art*, pp. 147–148.
60. Stocks, *Stoneworking Technology*, pp. 169–177.
61. Stocks, *Stoneworking Technology*, p. 176.

Bibliography

Arnold, Dieter. *Building in Egypt.* New York: Oxford University Press, 1991.

———. *The Pyramid of Senwosret I.* New York: Metropolitan Museum of Art, 1988.

Bernal, Martin. *Black Athena.* New Brunswick, NJ: Rutgers University Press, 1987.

Borchardt, Ludwig. "Der Bau" in Friedrich von Bissing, ed., *Das Re-Heiligtum des Konigs Ne-woser-re (Rathures), Vol. 1.* Berlin: A. Duncker, 1905.

Brier, Bob. "How The Pyramids Were Built," in *Archaeology,* Vol. 60, No. 3 (2007).

Brier, Bob and Ronald S. Wade. "The Use of Natron in Human Mummification: A Modern Experiment," in *Zeitschrift fur Agyptische Sprache,* Vol. 124 (1997), 89–100.

———. "Surgical Procedures During Ancient Egyptian Mummification," in *Zeitschrift fur Agyptische Sprache,* Vol. 126 (1999), 89–97.

Bui, Hui Duong, et al. "First Results of Structural Analysis of the Cheops Pyramid by Microgravity," in *Proceedings of the First International Symposium on the Application of Modern Technology to Archaeo-*

logical Exploration at the Giza Necropolis. Cairo: Egyptian Antiquities Organization Press, 1988, 66–90.

Burton, Richard, trans. *The Book of the Thousand Nights and a Night.* 6 vols. New York: Limited Edition Club, 1934.

Butler, Hadyn R. *Egyptian Pyramid Geometry.* Mississauga, Ontario: Benben Publications, 1998.

Clarke, Somers and R. Engelbach. *Ancient Egyptian Construction and Architecture.* Mineola, NY: Dover, 1990.

Cottrell, Leonard. *Mountains of Pharaoh.* New York: Rinehart & Co., 1956.

Dassault Systèmes. *Khéops Révélé: Revue de Presse.* Paris: Dassault Systèmes, 2007.

David, Rosalie. *The Pyramid Builders of Ancient Egypt.* London: Routledge, 1986.

Davidovits, Joseph and Margie Morris. *The Pyramids.* New York: Hippocrene Books, 1988.

Davidson, D. and H. Aldersmith. *The Great Pyramid and Its Divine Message.* London: Williams & Norgate, 1936.

Davidson, David. *The Great Pyramid's Prophecy Concerning the British Empire and America.* London: Covenant Publishing Co., 1932.

Diodorus of Sicily. *Library of History, Vol. I.* Cambridge, MA: Harvard University Press, 1968.

Dormion, Gilles, *Pyramide of Chéops: Architecture des Appartements.* Lille: self-published, 1996.

Drower, Margaret. *Flinders Petrie: A Life in Archaeology.* London: Victor Gollancz, 1985.

Edgar, John. *The Great Pyramid Passages and Chambers, Vol. I.* Glasgow: Bones & Hulley, 1910.

Edwards, I.E.S. *The Pyramids of Egypt.* Baltimore: Penguin Books, 1972.

Eissa, N. A., et al. 1988. "Application of Mossbauer and X-ray Fluorescence Spectroscopies to Study Sand and Limestone From: 1) Cavity in Passage to Queen Chamber in Khufu Pyramid. 2) Covers of Pits Closing Khufu Boats," in *Proceedings of the First International Symposium on the Application of Modern Technology to Archaeological Exploration at the Giza Necropolis.* Cairo: Egyptian Antiquities Organization Press, 1988, 233–247.

Evans, Humphry. *The Mystery of the Pyramids.* New York: Thomas Y. Crowell, 1986.

Fakhry, Ahmed. *The Monuments of Sneferu at Dashur.* 2 vols. Cairo: Government Printing Offices, 1959–61.

———.*The Pyramids.* Chicago: University of Chicago Press, 1961.

Gillings, Richard. *Mathematics in the Time of the Pharaohs.* New York: Dover, 1982.

Goidin, Jean Patrice and Gilles Dormion. 1988. "Architectural Analysis of the Horizontal Passage and the Discharge Chambres,"in *Proceedings of the First International Symposium on the Application of Modern Technology to Archaeological Exploration at the Giza Necropolis.* Cairo: Egyptian Antiquities Organization Press, 1988, 184–205.

Goneim, M. Zakariah. *The Buried Pyramid.* London: Longmans, 1956.

Greaves, John. *Pyramidographia.* London: J. Brindley, 1736.

Grinsell, L. V. *Egyptian Pyramids.* East Gate, Gloucestershire, UK: John Bellows, 1947.

Habachi, Labib. *The Obelisks of Egypt.* New York: Scribners, 1977.

Hart, George. *Pharaohs and Pyramids.* London: Herbert Press, 1991.

Hawass, Zahi, "The Pyramids and Temples of Egypt: An Update," in Petrie, W. M. Flinders, *The Pyramids and Temples of Gizeh.* London: Histories & Mysteries of Man, 1990.

———. *Mountains of the Pharaohs.* Cairo: American University of Cairo Press, 2006.

———. 2007. *Pyramids: Treasures, Mysteries, and New Discoveries in Egypt.* Vercelli, Italy: White Star.

Herodotus. *History.* Cambridge, MA: Harvard University Press, 1990.

Hodges, Peter. *How the Pyramids Were Built.* Oxford, UK: Aris & Phillips, 1993.

Houdin, Jean-Pierre. *Khufu.* Cairo: Farid Atiya Press, 2006.

Houdin, Jean-Pierre and Henri Houdin. 2002. "La Construction de la Pyramide de Khéops: Vers la Fin des Mystères?" in *Annales des Ponts et Chaussées,* Vol. 101 (2002). Paris: Ingénieur Science Société.

———. *La Pyramide de Khéops.* Paris: Editions Du Linteau, 2003.

Isler, Martin. *Sticks, Stones, & Shadows: Building the Egyptian Pyramids.* Norman, OK: University of Oklahoma Press, 2001.

Jackson, Kevin and Jonathan Stamp. *Building the Great Pyramid.* Toronto: Firefly Books, 2003.

James, T.G.H. *Pharaoh's People.* Chicago: University of Chicago Press, 1984.

Jecquier, Gustave. *La Pyramide d'Oudjebten.* Cairo: L'Organization Egyptienne Générale du Livre, 1978.

Jenkins, Nancy. *The Boat Beneath the Pyramid.* London: Thames & Hudson, 1980.

Kiner, Aline. "Révélations sur Khéops," in *Le Nouvel Observateur Science et Avenirs*, April 2007.

Knight, Charles. *The Mystery and Prophecy of the Great Pyramid.* San Jose, CA: Rosicrucian Press, 1928.

Krauss, Rolf. "The Length of Sneferu's Reign and How Long It Took to Build the Red Pyramid," in *Journal of Egyptian Archaeology*, Vol. 82 (1996), 43–50.

Landstrom, Björn. *Ships of the Pharaohs.* New York: Doubleday, 1970.

Lauer, Jean-Philippe. *Saqqara.* London: Thames & Hudson, 1976.

Lawton, Ian and Chris Ogilvie-Herald. *Giza: The Truth.* London: Virgin Books, 1999.

Lefkowitz, Mary R. and Guy MacClean Rogers, eds. *Black Athena Revisited.* Chapel Hill, NC: University of North Carolina Press, 1996.

Lehner, Mark, ed. *Giza Reports, Vol. I.* Boston: Ancient Egypt Research Associates, 2007.

———. *The Pyramid Tomb of Hetep-heres and the Satellite Pyramid of Khufu.* Mainz Am Rheim, Germany: Von Zabern, 1985.

———. *The Complete Pyramids.* London: Thames & Hudson, 1997.

Lemesurier, Peter. *The Great Pyramid Decoded.* New York: St. Martin's Press, 1977.

Lichtheim, Miriam. "Three Tales of Wonder," in *Ancient Egyptian Literature, Vol. I: The Old and Middle Kingdoms.* Berkeley: University of California Press, 1975.

Lipke, Paul. *The Royal Ship of Cheops.* Greenwich, UK: National Maritime Museum, 1984.

Lucas, A. *Ancient Egyptian Materials and Industries.* London: Histories & Mysteries of Man, 1989.

Macaulay, David. *Pyramid.* Boston: Houghton Mifflin Co., 1975.

Manetho. *Aegyptiaca.* Cambridge, MA: Harvard University Press, 1971.

Maragioglio, V. and C. Rinaldio. *L'Architettura Delle Piramidi Menfite*, Vols. I–VII. Rapallo: Officine Grafiche Canessa, 1963–1977.

McCarty, Louis. *The Great Pyramid of Jeezeh.* San Francisco: self-published, 1907.

Mendelssohn, Kurt. *The Riddle of the Pyramids.* New York: Praeger, 1974.

Naville, Édouard. *The Temple of Deir el Bahri, Part VI.* London: Egypt Exploration Fund, 1908.

Newberry, Percy E. *El Bersheh, Part I.* London: Egypt Exploration Fund, 1895.

Ostrander, Sheila and Lynn Schroeder. *Psychic Discoveries Behind the Iron Curtain.* Englewood Cliffs, NJ: Prentice Hall, 1970.

Parry, Dick. *Engineering the Pyramids.* Thrupp, Gloucestershire, UK: Sutton Publishing Ltd, 2004.

Perring, J. E. *The Pyramids of Gizeh from Actual Survey and Admeasurement.* London: James Fraser, 1839.

Petrie, W. M. Flinders. *The Pyramids and Temples of Gizeh.* London: Histories & Mysteries of Man, 1990 (reprint).

———. *Researches in Sinai.* London: John Murray, 1906.

Romer, John. *The Great Pyramid.* Cambridge, UK: Cambridge University Press, 2007.

Rowbottom, William. *The Mystery of the Bible Dates Solved by the Great Pyramid.* London: W. H. Guest, 1877.

Rutherford, Adam. *A New Revelation in the Great Pyramid.* London: Institute of Pyramidology, 1945.

———. *The Great Pyramid.* London: self-published, 1939.

Scamuzzi, Ernesto. *Egyptian Art in the Egyptian Museum of Turin.* New York: Harry N. Abrams, 1965.

Seiss, Joseph A. *A Miracle in Stone: The Great Pyramid of Egypt.* Philadelphia: Porter & Coates, 1877.

Smith, Craig B. *How the Great Pyramid Was Built.* Washington D.C.: Smithsonian Press, 2004.

Smith, Joseph Lindon. *Tombs, Temples, & Ancient Art.* Norman, OK: University of Oklahoma Press, 1956.

Smyth, Piazzi. *The Great Pyramid: Its Secrets and Mysteries Revealed.* New York: Bell Publishing, 1975 (reprint).

Spence, Kate. "Ancient Egyptian Chronology and the Astronomical Orientation of Pyramids," in *Nature*, Vol. 408 (2000), 320–324.

Stocks, Denys. *Experiments in Egyptian Archaeology.* London: Routledge, 2003.

Tompkins, Peter. *Secrets of the Great Pyramid.* New York: Harper & Row, 1971.

Tonouchi, Shoji. "Non-Destructive Pyramid Investigations and Physical Property of the Sand Found Inside Pyramid" in *Proceedings of the First International Symposium on the Application of Modern Technology to Archaeological Exploration at the Giza Necropolis*. Cairo: Egyptian Antiquities Organization Press, 1988, 91–105.

Verner, Miroslav. *Forgotten Pharaohs, Lost Pyramids, Abusir.* Prague: Academia Skodaexport, 1994.

———. *Pyramids.* New York: Grove Press, 2001.

Vyse, Howard. *Operations Carried on at the Pyramids of Gizeh in 1837.* 3 vols. London: James Fraser, 1837–42.

Photography Credits

All images courtesy of Bob Brier except:

Images throughout text:

Jon Bodsworth: pages 16-7, 31, 82-3 (bottom).

Jean-Pierre Houdin: pages 39, 83, 86, 95 (right), 103, 106, 143.

Val Parks: page 56.

Albert Ranson: pages 62-3, 129.

Farid Atiya: page 74.

Dassault Systèmes: pages 98-9, 101, 105, 109, 112, 123 (bottom), 131, 132, 144, 146-7, 148, 159, 164-5, 169.

Edgar Brothers: page 104.

Henri Houdin: pages 127, 128.

EDF Foundation: page 134.

Images in color inserts:

Jon Bodsworth: page 1.

Hulton-Deutsch Collection/Corbis: page 2.

Dassault Systèmes: pages 3, 4, 5, 6, 7, 8.

Index